OXFORD TEXTBOOK OF FUNCTIONAL ANATOMY

VOLUME 3

OXFORD TEXTBOOK OF FUNCTIONAL ANATOMY

VOLUME 3

Head and Neck

by PAMELA C. B. MacKINNON
and JOHN F. MORRIS

Department of Human Anatomy, Oxford University

OXFORD UNIVERSITY PRESS

Oxford University Press, Walton Street, Oxford OX2 6DP

Oxford New York
Athens Auckland Bangkok Bombay
Calcutta Cape Town Dar es Salaam Delhi
Florence Hong Kong Istanbul Karachi
Kuala Lumpur Madras Madrid Melbourne
Mexico City Nairobi Paris Singapore
Taipei Tokyo Toronto
and associated companies in
Berlin Ibadan

Oxford is a trade mark of Oxford University Press

Published in the United States by
Oxford University Press Inc., New York

British Library Cataloguing in Publication Data
MacKinnon, Pamela C. B.
Oxford textbook of functional anatomy.
Vol. 3, Head and neck
I. Title II. Morris, John F.
611
ISBN 0 19 261519 X

Library of Congress Cataloging in Publication Data
(Revised for vol. 3)
MacKinnon, Pamela C. B.
Oxford textbook of functional anatomy.
(v. 1: Oxford medical publications)
Includes indexes.
Contents: v. 1. Musculoskeletal system—v. 2.
Thorax and abdomen—v. 3. Head and neck.
1. Human anatomy. 1. Morris, John F. II. Series.
[DNLM: 1. Anatomy. QS 4 M1580]
QM5.M33   1986   611   85-7099
ISBN 0 19 261517 3 (pbk. : v. 1)
ISBN 0 19 261519 X (v. 3)

Printed and bound in Hong Kong

Acknowledgements

We are greatly indebted to Audrey Besterman, who drew the artwork, for both her artistic skill and her knowledge and understanding of anatomy. We also wish to thank Drs Basil Shepstone and Stephen Golding, Department of Radiology, Oxford, who wrote Chapter 3, 'Medical imaging in the head and neck,' provided many of the other radiographs, and advised on the radiology sections throughout the book. Similarly, we thank Dr Gillian Morriss-Kay, Department of Human Anatomy, Oxford, who wrote Chapter 5, 'Development of the head and neck', advised on the embryology interspersed through the text, and provided Fig. 6.1.1. We are also grateful to the many other clinical and preclinical colleagues who generously provided illustrative material and helpful advice on specific sections: Drs Philip Anslow and Andrew Molyneux for CTs of the head; Dr Gordon Ardren for the radiographic images of swallowing (Fig. 6.5.8); Dr Christopher Burke for Fig. 6.8.3; Mr Andrew Freeland, FRCS and Mr William Lund, FRCS for the sections relating to ear, nose, and throat and Figs 6.6.2 and 6.9.4; Mr John Raine, FRCS for applied anatomy of the facio-maxillary region and Fig. 6.3.12; Miss Peggy Frith, FRCS for Fig. 6.8.5a; Dr David Hilton-Jones for Figs 6.12.4, 6.13.4, and 6.14.4; Dr Roy Kay for advice on the auditory system; Mr Michael Poole, FRCS for Figs 6.1.4 and 6.1.6; Dr Kathy Sulik for Fig. 5.5; Dr Jeffrey Wright for Fig. 6.8.5b.

We also gratefully acknowledge the help of Brian Archer and the photographic unit of the Department of Human Anatomy, Oxford for many of the photographic illustrations; also Molly Harwood and Jane Thomas in the Department of Medical Illustration at the John Radcliffe Hospital, Oxford; Herbert Merry for allowing us to publish Fig. 6.1.5; the staff of the Oxford University Press who have seen the project through from beginning to end; and last, but not least, Roger White and Terence Richards for carefully conserving prosected material in the Department of Human Anatomy of which we have made consistent use.

We remember, too, the instruction of our own mentors and the enthusiastic support of our professional colleagues. To all of these and, in particular, Professors Ruth Bowden, the late Sir Wilfrid LeGros Clark, FRS, Sir Barry Cross, FRS, the late Geoffrey Harris, FRS, Charles Phillips, FRS, and Joseph Yoffey, our warm and grateful thanks.

Contents

CHAPTER 1

General introduction

Expectations and use of the book

In writing this book we have deviated from traditional approaches and attempted to emphasize functional and living anatomy and its imaging in living subjects. This book, like Volume 1, *Musculoskeletal system*, and Volume 2, *Thorax and abdomen*, has been designed to enable students of morphology, whether they be medical or dental, physiotherapists or radiographers, nurses, or any other student of the biology of Man, to understand the functional design of their own body and that of others. To dissect an entire cadaver can be very instructive and is essential for aspiring surgeons. However, the head and neck is a particularly difficult region to dissect successfully. Also, the recent escalation of knowledge in all branches of the medical sciences, for which time must be provided in teaching programmes, makes very extensive dissection impractical in most courses. Nevertheless, the understanding of how the body is built and how it works is fundamental to the practice of all aspects of medicine. This volume therefore guides its readers through fundamentals of the anatomy of the head and neck by examination of the living body and by use of appropriate prosections. For those who do not have access to prosected and dissectable material, the book is extensively illustrated. The embryological development of the head and neck (excluding the brain) has been outlined, where appropriate, to aid understanding both of normal anatomy and of certain abnormalities which will be encountered. To achieve the very necessary integrated view of the functional anatomy of the head and neck, liberal reference should be made to embryology, histology, neuroanatomy, physiology, and biochemistry textbooks. It is important that you are able to answer the questions which have been inserted throughout the book; if the foregoing text has been absorbed and understood, they should not prove difficult. Personal notes on your anatomical investigations made in the margins of the book will prove invaluable. At some future time, rapid perusal of a seminar containing additional comments of your own will ensure an equally rapid recall of the basic information. This is a book which, if used properly, should not be discarded.

Functional anatomy

Functional or living anatomy is the study of structure and function of the body in its living state; of the skeletal system which supports, protects vital organs, and provides levers to which muscles are attached; of muscles and joints which provide for movement between the various skeletal units; of the highly specialized cardiovascular system through which oxygen and nutrients are pumped to individual cells of the body and waste materials are taken for excretion; of the various organs in the head and neck, thorax, and abdomen which enable the body to remain viable by ensuring a homoeostatic environment for each individual cell; of the reproductive system which ensures continuity of the species; and of the nervous system which receives and integrates information from both the internal and external environments and which, through speech, movements, and behaviour, enables us to interact with our environment and express our individual character and personality. This volume concerns the head and neck and is set out in 'seminars' each of which require two or more hours of study. Seminars 11–14 on the cranial nerves can conveniently be studied alongside a course in neuroanatomy while Seminars 1–6 and 10, which deal with the upper respiratory tract and its related structures. can precede study of the thorax (Vol. 2) if so desired.

The changing form of the body and its relation to function

Growth and development of the body depend on intracellular mechanisms, the coding of which is unique for any one person, with the exception of homozygous twins. The body develops in the uterus over a period of about 40 weeks and this development usually produces a 'normal' newborn individual. The body develops and grows further during childhood and adolescence to attain adult form. Occasionally, some part of this development is imperfect to a greater or lesser degree.

Certainly, great variation among normal individuals exists and it is important that you develop a concept of this **range of normality** so that you can judge what is abnormal. For this purpose some illustrations of abnormalities are included in the book. In later adult life, ageing changes lead to senescence. Remember that most bodies donated for examination in the dissecting rooms are those of elderly people.

External differences between male and female are mostly obvious; however, the body of both sexes undergoes many cyclic changes of a circhoral and diurnal nature and that of the female undergoes, in addition, a monthly cycle. Throughout life, the body responds morphologically to functional demands (e.g. muscle hypertrophy as a result of exercise, goitres of the thyroid if the diet is deficient in iodide) and to abuses and injuries by repair and healing.

'The body' which you must consider is therefore not the static, usually elderly form which you see on the dissecting room table, but rather a living, dynamic organism, constantly changing and responding to the functional challenges of its environment.

Any region of the body consists of a number of different tissues. The anatomical form that these tissues take in any region is the result of their evolution to fulfil a functional role. The structure of tissues can be considered on two levels: microscopic structure is the object of study in histology; macroscopic (naked eye) and radiological structure is the subject of these seminars.

Terms used in anatomical description (1.1a)

For ease of communication and convenience of description the body is always considered as standing erect, facing ahead, the arms by the sides, and the palms of the hands facing forward with the fingers extended. Place yourself in this position and note that it differs in a number of ways from your normal standing posture.

The terms **anterior** (ventral) and **posterior** (dorsal) refer to structures towards the front and back of the body respectively. Structures in the anteroposterior (A–P) midline are said to be **median**, those close to the midline **medial**, and those further away **lateral**. Structures above are referred to as **superior** (or cephalic), structures below as **inferior** (or caudal). **Proximal** means nearer to the origin of a structure, **distal** is the opposite. **Superficial** means nearer the skin, **deep** is the opposite.

Anatomical planes (1.1b)

Sagittal — a vertical anteroposterior plane (longitudinal).
Coronal — a vertical plane at right angles to the sagittal planes.
Transverse — a horizontal plane at right angles to both coronal and sagittal planes.
Oblique — any plane that is not coronal, sagittal, or transverse.

Movements (1.2)

A forward or anterior movement of the head as in nodding is usually termed **flexion**; a backward or posterior movement is **extension**. If the forward-facing (non-rotated) head is kept straight and the neck is bent to either side the movement is referred to as **lateral flexion**. Rotation of the head occurs on the axis of the spine. Specific movements of chewing, swallowing, phonation, facial movements, and movements of the eye and pupil are dealt with in the appropriate seminars.

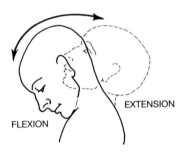

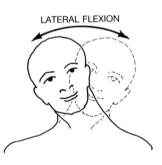

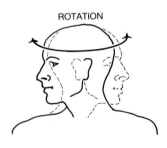

1.2
Anatomical terms used to describe movements of head and neck.

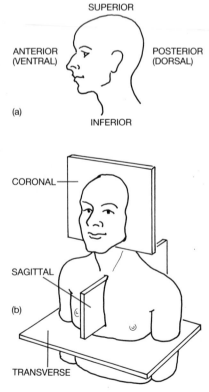

1.1
Anatomical terms used in description of head and neck.

CHAPTER 2

Systematic thinking about tissues

This chapter outlines the range of topics that should be considered when studying tissues of the head and neck.

Skin

Consider:

• The **degree of keratinization**: keratinization is protective but makes the skin less flexible. Compare, for example, the skin of the forehead or back of neck with that of the lips or eyelids. The degree of keratinization can be altered by functional demands: for instance, the callus on the lip of a professional trumpeter.

• The **degree of hairiness** and type of hair: only the palms, soles, eyelids (not the eyelash area), and penis are hairless; however, in older men a receding hair line on either side of the forehead and an extending bald patch in the centre of the scalp become increasingly likely. The **density** and **coarseness** of hair differs from region to region and between sexes (e.g. coarse vibrissae within the external auditory meatus and the nostrils). With increasing age, in both sexes, the hair tends to lose its curl and becomes wispy, losing colour until it turns grey and eventually white. The **distribution** of body hair is sexually dimorphic; in the male (but not the female) abundant hair is present over the side of the cheeks ('sideboards'), jaw (beard), upper lip (moustache), and the midline of the abdomen from the pubis to the umbilicus. This distribution is a secondary sexual characteristic which develops at puberty in response to increasing plasma androgens. Abnormal hair distribution can reflect endocrine imbalance.

• The presence of **dermal ridges**: these are present on the palms and soles of the lands and feet of Man. They function in gripping and also in texture recognition since, when the ridged skin is moved over an object of interest, vibrations are produced which can be sensed and interpreted. The pattern of ridges on the pads of digits forms the fingerprint which, except in homozygous twins, is unique for each individual.

• The **sweat glands** and their openings: in Man openings of **eccrine sweat glands** are located over most of the body surface. They are concentrated on the finger pads where their ducts open on to the dermal ridges. Eccrine sweat glands secrete a clear watery saline and are important in the control of body temperature; the number of open, active sweat glands on most parts of the body is increased by exercise. However, those on the palms, soles, and sometimes the forehead and upper lip also respond to changes in emotional status (it was said that Chinese jade traders paid great attention to the extent of sweating on the upper lip of a potential buyer when negotiating a price). **Apocrine sweat glands**, found in the axilla and external genitalia, produce sweat with a higher organic content due to extrusion of their cellular content. Odour formation is due to breakdown of this organic content by bacterial enzymes on the skin surface.

• The **sebaceous glands**: these open on to hair follicles and are numerous on the scalp and around the nose. Their secretion is derived from whole cells which are sloughed off into the lumen of the gland to provide an oily protective covering for the hair.

• The **skin creases**: these pronounced lines, often prominent on the forehead, face, and around joints, occur where skin is firmly attached to the underlying tissues. They become more marked with age due to the habitual patterns of movement, for example, of facial skin. Such **flexure creases** differ from the fine crease-like lines which appear in the skin of old people and which are caused by degeneration of collagen fibres and reduced attachment of the skin to underlying tissues.

• The **skin cleavage lines** (of Langer): these cannot be seen, but maps are available; in general, they run circumferentially round the neck and trunk. They mark the orientation of transverse fibrous tissue bundles in the skin. An incision **along** a Langer's line will heal with a fine, barely noticeable scar; for example, the transverse incision used to approach the thyroid gland, which can be virtually hidden in a skin crease. Incisions **across** the lines tend to produce more pronounced scars because of the greater amount of fibrous tissue laid down.

• The **blood supply of the skin**: this is derived from local vessels. Capillary loops can be observed if the skin of the nail bed is cleared with a drop of oil and observed with a dissecting microscope. The thermal sensitivity of skin vasculature can readily be demonstrated with hot and iced water. In cold conditions much of the cutaneous blood supply is short-circuited from arterioles to venules by arteriovenous anastomoses.

• The **innervation of the skin**: stimuli to the skin are defined in terms of forms of energy (e.g. mechanical or thermal), their spatial distribution, intensity, and rate of change. Some cutaneous sense organs are highly sensitive to mechanical, others to thermal and pain-producing stimuli; some adapt rapidly, others more slowly. Variations in mechanoreceptive spatial discriminatory power are measured by the two-point discrimination threshold, the smallest distance between two simultaneously applied mechanical stimuli that can be perceived as two rather than one. Temperature receptors are densely distributed over the face; two-point discrimination is very acute on the tongue. Babies and young children, as well as animals, explore objects orally.

You will need to know:

● The **local nerve of supply** to an area of skin: this is required both for diagnosis of nerve lesions and for application of local anaesthesia.

● The **spinal** or **cranial nerve** which supplies the area of skin: this is required for the diagnosis of spinal cord and spinal nerve lesions on the one hand and brainstem and cranial nerve lesions on the other. The area of skin supplied by a spinal nerve is called a **dermatome**.

Superficial fascia

This is the subcutaneous connective tissue. It varies considerably from place to place. Over the face, muscles of facial expression run between the facial bones and skin of the face so that no distinction between superficial and deep fascia is possible.

Consider:

● The degree of **fat** accumulation and its regional variation: compare the accumulation of fat in the neck of some older people with lack of such accumulation on the tip of the nose or the pinna of the ear.

● The **fibrous tissue** content: this determines the attachment of the skin to deeper structures. The skin of the scalp is very tightly attached to an underlying aponeurosis which unites pairs of subcutaneous muscles over the forehead and occiput. Compare this with the loose fascia deep to the skin of the eyelids. Fibrous tissue is prominent at skin creases on the forehead and face.

Deep fascia

This is the connective tissue that covers and ensheathes muscles and helps to attach them to bones. However, the muscles of the face (and certain other areas) are attached between bone and skin so that no distinction between superficial and deep fascia can be made. Deep fascia is mainly fibrous, but also contains fat and fluid. Its thickness varies very considerably depending on functional requirements. Over expansile regions such as the pharynx or buccinator it is very thin; in the neck, however, **sheets** of deep fascia form non-expansile sleeves which invest groups of muscles in different layers, thus enabling them to move relative to each other with the minimum of friction. In areas of the body such as the thigh, non-expansile sleeves of fibrous tissue play an important part in the mechanics of venous return. Where sheets of strong deep fascia exist, they often form extra sites for muscle attachments.

Bones

You should be able to identify the bones of the face, skull, and neck. The skull has sexually dimorphic features but they are less obvious and less reliable discriminators than those of the pelvis (Vol. 2, p. 105).

Consider:

● The overall **shape** of the bone.

● The position and shape of **articular surfaces** (e.g. on the head of the mandible or the ossicles of the middle ear) and the structures with which they articulate.

● **Named parts** and **prominences**, especially those that are palpable in the living subject.

● The site of major **muscle attachments** which may or may not be associated with roughened areas or protrusions of the bone.

● The site of major **ligament and membrane attachments**.

● The **blood supply** and position of any nutrient arteries; **exit foramina** for draining veins.

● The **marrow cavity** content of red or fatty marrow; its extent in both the child and the adult.

● Any specializations of the **trabecular** pattern within the bone or thickenings of the cortex which reinforce particular lines of stress in the bone.

● The **ossification** of the bone — and whether it occurs in **membrane** or in **cartilage**. Bones of the base of the skull and cervical vertebrae ossify in cartilaginous models, whereas most bones of the vault of the skull, face, and the clavicle ossify in membranous models. Their **primary centres of ossification** appear *in utero*, and any **secondary centres** around puberty, fusing between 17 and 25 years. The time of appearance of these centres need not be committed to memory, but you should acquire a general understanding of the sequence of ossification of the different parts. The development of the different centres can be useful in determining the age of bones, and in assessing whether the skeletal age of a child matches its chronological age.

Joints (see also Vol. 1, Chapter 2)

Joints are the articulations between bones. Their form and structure vary widely in relation to the functional requirements of the articulation and, in any joint, reflect an evolutionary compromise between mobility and stability. The degree of **mobility** varies widely: some joints essentially permit no movement (e.g. the teeth in the jaw, or the sutures of the adult skull); in others, small gliding or angular movements occur (e.g. at articular facets between cervical vertebrae); in yet others (e.g. the temporomandibular joint), more extensive movement in different planes can occur. All joints must have sufficient **stability** to resist dislocation by the forces to which they are normally subjected.

Joints can be classified in various ways based on criteria such as: the type of tissue that separates/unites the bones; the extent or type of movement that occurs; the stability of the joint; the temporary or permanent nature of the joint. They are commonly classified into different histological types depending on the tissues that separate or unite the bones (fibrous tissue, hyaline cartilage, synovial fluid).

The movements that can occur depend on the actions and power of muscles crossing the joint; the extent to which tissues that unite the bones can be deformed (partly determined by their length and cross-sectional area); the shape of the articulating surfaces; the ligaments that surround the joint; and any soft tissue appositions that occur as a result of movement. In addition to producing one particular movement of a joint, a muscle can also prevent occurrence of (antagonize) the opposite movement, thus providing active stabilizing support for the joint. This is particularly important at joints through which there is transmission of considerable force. Indeed, the active support that can stabilize a joint in any position is provided only by the simultaneous contraction of antagonistic pairs of muscles. The ligaments resist movement only at the **limits** of the normal range. Should the muscles become paralysed then, depending on the forces tending to deform the joint, ligaments will stretch and permanent joint deformity may ensue.

Where solid fibrous tissue or hyaline cartilage link the bones (**fibrous** or **'primary' cartilaginous joints**, respectively), the amount of movement will depend on the degree to which that tissue can be deformed by muscular effort or external forces. Compare the lack of movement and great stability of sutures (fibrous joints) in the adult skull, with the deformability of the fetal skull during birth. In the fetus, bones of the skull vault are widely separated by thin sutural fibrous tissue; in adults the bones, which often interlock, are united by thick fibrous tissue. In **'secondary' cartilaginous joint** such as the intervertebral discs, the articulating bony surfaces are covered by hyaline cartilage and linked together by fibrocartilage, which may have a significant fluid component. Each intervertebral disc permits only a small degree of movement, but the sum of such movements gives the cervical spine considerable flexibility. As discs age, fluid is lost, and mobility is reduced. In **synovial joints**, the hyaline cartilage-covered articular surfaces of bones (or cartilages, as in the larynx) are separated by a joint cavity filled with synovial fluid and surrounded by a capsule lined with synovial membrane. Such joints are potentially quite freely movable (e.g. temporomandibular and atlanto-occipital joints). However, at other synovial joints movement is negligible (sacroiliac joint; Vol. 1, Chapter 7, Seminar 1). Further classification of synovial joints is based on the shape of the articular surfaces: plane; hinge; pivot; condylar; ellipsoid; saddle; ball-and-socket. These descriptions are, of course, approximate. The shape of the articular surfaces (together with the ligaments) determine the planes and axes of movement that are possible.

For any joint consider:

- The extent and type of **movement** that occurs at the joint.
- The **stability** of the joint.
- The type of **tissue** that separates the bones and its physical dimensions.
- The **articulating surfaces**: the bones that take part; the shape of their articular surfaces; the areas of contact in different movements.
- The **capsule**: its extent, attachments, strengths, and deficiencies. The capsule is usually attached to the margins of the articular surfaces of a synovial joint.
- The **ligaments**:
  (a) **Intrinsic** ligaments are thickenings of the capsule along particular lines of stress.
  (b) **Accessory** ligaments also limit movement of the joint in various directions but are separated from the capsule.
- Any **synovial cavity and membrane**: synovial membrane usually lines all non-articular surfaces within a joint. It secretes synovial fluid which lubricates and, in part, provides nutrition for the joint. The amount of synovial fluid is very small. On radiographs of synovial joints, the 'space' between the articulating bony surfaces is occupied almost entirely by the radiolucent articular cartilage; the fluid 'space' is minimal in most joints.
- **Intra-articular discs** of fibrocartilage usually divide the cavity of joints in which movement occurs in two separate axes (e.g. temporomandibular joint).
- **Blood supply**: around joints there is usually an anastomosis derived from local arteries, which give branches to the capsule.
- **Nerve supply**:
  (a) The capsule of joints has an important **sensory** nerve supply that conveys **mechanoceptive** information to the central nervous system concerning movement of the joint, and **pain** fibres which signal excessive movement and thus protect a joint.
  (b) **Vasomotor** fibres of the sympathetic nervous system supply arterioles of the synovial membrane. A nerve which supplies a muscle acting on a given joint will also supply sensory fibres to the capsule of that joint.

Skeletal muscles and the movements they produce

Muscles acting in any movement may be classified as: **prime movers** (agonists) which produce the required movement; **synergists** which act with the prime mover to produce the desired movement; **antagonists** which oppose the prime mover; **essential fixators** which permit movement from a stable base; **postural fixators** (e.g. of the trunk) which prevent the body being toppled by movements which shift the centre of gravity. 'Paradoxical' actions counter the force of gravity (e.g. spinal extensors contract when a heavy weight is lowered to the ground by flexion of the spine). Some programmes of muscle action are relatively stereotyped; others (e.g. for speech) have to be learned. A muscle may be a prime mover in one programme, an antagonist in another, a synergist or essential fixator in yet others.

Many details of muscle topography are clinically unimportant, but consider:

● The **attachments** of the muscle: most muscles are attached to separate bones and cross joints; superficial muscles of the face, however, pass from bones to skin. The attachment that usually remains fixed during the prime movement is the **origin** of a muscle; the **insertion** is the attachment to the part that moves.

● The **shape** of a muscle and **arrangement of its fibres**: the degree of shortening is proportional to the length of the muscle fibres; the power of a muscle is proportional to the number of muscle fibres.

● The **blood supply** to muscles: muscles need a good blood supply, but details of arteries which supply them are rarely important.

● The **nerve supply**: muscles of the head and neck receive motor and sensory supplies from either cranial or cervical spinal nerves; postganglionic sympathetic fibres join these nerves for distribution to blood vessels in the muscle.

Visceral muscle

Visceral (unstriated, smooth) muscle is found in many different parts of the head and neck. These include arterioles which, because smooth muscle can contract in a sustained manner, regulate blood flow to the tissues and organs; the pupil and lens of the eye, to alter their diameter and curvature, respectively; and the ducts of lacrimal and salivary glands. Visceral muscle is innervated by autonomic nerves.

Arteries

For any artery consider:

● The commencement and parent vessel.

● The **course**, particularly points at which the vessel's pulsations are palpable (over bone) or the vessel is exposed to injury. Sites where arteries cross bone may form useful 'pressure points' where digital pressure can arrest haemorrhage (e.g. the facial artery passing over the body of the mandible, or the superficial temporal artery passing over the zygomatic process).

● The **major branches** and **mode of termination**.

● The **area of supply**.

● The **degree of anastomosis** with other major vessels. Where movements occur anastomoses are common. However, some vessels, such as the central artery of the retina, are truly '**end arteries**', with no anastomosing vessels; many are functionally end arteries because of very limited anastomosis, e.g. the internal carotid arteries and the vertebral arteries anastomose at the base of the brain to form the 'circle of Willis'. In children this union can compensate for major changes in flow, but this is rarely the case in adults.

● Availability of the vessel for puncture and introduction of a catheter, as for angiography.

Veins and venous sinuses

Consider:

● The **mode of commencement** and the **area drained**.

● The position: **superficial** (in superficial fascia); or **deep** (beneath the deep fascia).

● The **course**: particularly where the vein can be punctured with a needle for intravenous administration of substances or withdrawal of blood.

● The extent of any **valves** within the veins. There is a marked regional variation with regard to the incidence of valves. Veins of the head and neck mostly lack valves. Negative intrathoracic pressure generated by respiration, and the effect of gravity both facilitate the return of venous blood to the heart.

● Any **major tributaries**.

● The **degree of anastomosis** with other veins: for instance between veins draining the scalp and extradural veins within the cranial cavity.

● The **mode of termination**.

● **Portal veins**: on the stalk of the pituitary gland at the base of the brain is a set of portal veins, with capillaries at both ends, which link the hypothalamus and anterior pituitary. This small but important portal system conveys hormones synthesized by hypothalamic neurons to the anterior pituitary, where they regulate the synthesis and secretion of its various hormones.

Lymphatics

Lymphatic vessels are difficult to dissect and cannot be detected by routine examination of a living subject unless they are inflamed. They can, however, be demonstrated radiologically after injection of radio-opaque dyes. You must become familiar with the lymphatic drainage of an area since not only infection but also malignant tumours can spread by this route. There is no lymphatic drainage of the central nervous system, or the cornea.

Consider:

● The **area drained** by the lymph vessel or node

• Its position: **superficial** lymph vessels draining the superficial aspects of the scalp, face, and neck lie in superficial fascia close to veins and drain into groups of **superficial lymph nodes** which form a collar around the upper part of the neck. From there, they drain along superficial lymph trunks to **deep lymph nodes** which lie vertically alongside the principal deep blood vessels of the neck. These deep lymph nodes also drain **deep** lymphatic vessels of the tongue, pharynx, and neck and a ring of deep lymphatic tissue which lies beneath the mucous membrane around the upper part of the pharynx, posterior nares, and tonsil.

• The **extent of anastomosis**: in general, there is considerable anastomosis between lymphatics serving adjacent areas. Thus, when lymphatics are blocked by tumour, nodes not normally draining an area may become involved.

• The **route** whereby lymph returns to the bloodstream. The main collecting lymph ducts of the head and neck drain into the junction between the subclavian and jugular veins on each side.

Nerves

For any nerve that you study consider:

• The **type of fibre** it contains:

(a) **Somatic** — serving the skin and musculoskeletal system. Individual fibres are either **motor** (efferent) to the skeletal muscle, or **sensory** (afferent) from skin, muscles, joints, and connective tissues.

(b) **Autonomic** — serving the eyeball, glands, viscera, and blood vessels. The autonomic system is subdivided into two efferent components: the **sympathetic** and **parasympathetic** systems, which differ in their origin, distribution, and function. In addition, **sensory** (afferent fibres) from the viscera and specialized sense organs of the blood vessels run with the autonomic nerves.

• The **origin** of the various fibres. **Somatic** innervation to tissues of the head and neck is derived from cranial nerves III–XII which take origin from the brainstem or from cervical spinal nerves C1–C5. In addition, the olfactory (I) and optic (II) and vestibulocochlear (VIII) cranial nerves supply the special senses. **Sympathetic** innervation to the head and neck originates from preganglionic neurons in the upper thoracic spinal cord, and is relayed by neurons of the cervical sympathetic ganglia. **Parasympathetic** innervation originates from the oculomotor (III), facial (VII), glossopharyngeal (IX), and vagus (X) cranial nerves and is relayed by the **parasympathetic ganglia** of the head (ciliary, pterygopalatine, submandibular, and otic) and by ganglion cells in the walls of the alimentary and respiratory tracts. Sensory innervation from these tracts and from the specialized arterial receptors passes with cranial nerves VII, IX, and X to the brainstem (solitary nucleus and tract).

• The **course** of the nerve, particularly where it can be palpated, and whether it is liable to trauma.

• The **major branches** of the nerve.

• The **muscles supplied** and the effect on movement of damage to the nerve.

• The area of **skin supplied**.

• The **glands supplied** (autonomic innervation) and the effects of the innervation on their function.

Glands

A number of **exocrine** (lacrimal, salivary, sweat) and **endocrine** (pituitary, thyroid, parathyroid) glands are located in the head and neck.

For both types of gland consider:

• The **function** of the gland and the nature of its **secretion**.

• The **position** of the gland and its immediate relations.

• The **innervation** of the gland and the effects of the innervation on secretion.

• The **blood supply** to the gland and (especially for endocrine glands) its venous drainage.

For exocrine glands also consider:

• The **duct** draining the gland, its termination, and any narrow points.

CHAPTER 3

Medical imaging in the head and neck

B. SHEPSTONE and S. GOLDING

One of the most important adjuncts to physical examination in the study of the anatomy of living subjects is **medical imaging**. Since the turn of the century images obtained with X-rays (**radiographs**) have been the standard method of visualizing many areas of the body. Images can now be obtained using sound waves (**ultrasound imaging**), or by using radiation emitted from substances which have been administered to the patient (**nuclear imaging**). Recently, computerized techniques have been developed which display a cross-sectional 'slice' or tomogram of the body. **Transmission computed tomography** (TCT, but usually abbreviated to CT) uses X-rays, whereas **emission computed tomography** (ECT) uses radionuclides. These types of computed tomography must be distinguished from the older **planar tomography** in which a computer is not used. In the case of ECT, there is a further subdivision into single-photon ECT (SPECT) and positron ECT (PET). (The acronyms are still in a state of flux with SPECT now being abbreviated to SPET (cf. PET, in which the C has always been omitted).) Computer techniques are also applied to the nuclear magnetic resonance effect to produce a **magnetic resonance image** (MRI) based on magnetic fields.

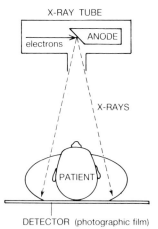

3.1
The principle of radiography.

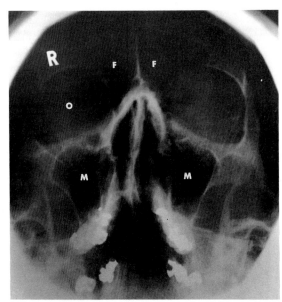

3.2
Frontal radiograph of bones of face.

Radiography

X-rays are part of the electromagnetic radiation spectrum and can be produced by bombarding a tungsten anode with electrons under high voltages. When the electrons strike the anode their kinetic energy is converted to heat and radiation, including X-rays. In medical radiography the tungsten anode is suspended over the patient so that the beam of X-rays passes through the body; the emerging radiation is then picked up by a detector which is usually photographic film (**3.1**). Since photographic film is sensitive to X-rays, the film will be exposed to a degree that depends on how much of the beam has passed through the patient and how much has been absorbed by the different tissues of the body. Air is radiolucent; bone and metal radio-opaque.

In the face, good contrast is provided by the bone, soft tissues, and air-containing sinuses and a clear image of the maxillary (M) and frontal sinuses (F) and orbits (O) is obtained (**3.2**). In the neck the air in the pharynx (P), larynx (L), and trachea (T) can be distinguished and the

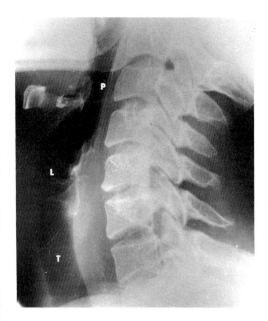

3.3
Lateral radiograph of neck.

spine is seen, but structures such as the oesophagus and blood vessels are not seen because they are of similar density to their surroundings (**3.3**).

Fluoroscopy ('screening')

If a fluorescent screen is substituted for photographic film a direct image is produced. This enables movements of organs to be studied. It is customary nowadays to view the image on a television screen; an amplifier system or image-intensifier is usually employed which ensures a good picture with a reduction in the dose of radiation. The fluoroscopic image can then be recorded by photography or video, or by a digital computer which stores the information received (**digital radiography**).

Planar tomography

This can be used to overcome the superimposition effects present in the conventional radiographic image. The X-ray tube and the film move around the patient in a constant relationship to the plane of interest which therefore appears as a sharp image; the overlying and underlying areas move relative to the tube and film so that their image is blurred. The effect of tomography is therefore to produce a clear image of one plane only. In the radiograph of the neck (**3.4**) the larynx is projected over the spine and is not clearly seen. On the tomogram (**3.5**) the plane of focus is such that only the larynx is seen and the vocal folds (arrow) and pyriform fossa (open arrow) are evident.

pension of barium sulphate is given to a patient by mouth. The pharynx can also be studied by a rapid series of radiographs taken during swallowing. In **3.6** a lateral radiograph shows barium distending the pharynx (P) and oesophagus (O) and causing the epiglottis to be inverted (arrow). Note that the soft palate and uvula close the nasopharynx and prevent reflux into the nose (open arrow).

Iodine-containing contrast media have many applications; they can be injected intravenously to be excreted by the kidneys (**intravenous urogram**). It is also possible to study blood vessels by inserting a fine tube (catheter) into an accessible vessel, for example the femoral artery, and passing its tip into the desired vessel, where contrast media can be injected. In **3.7**

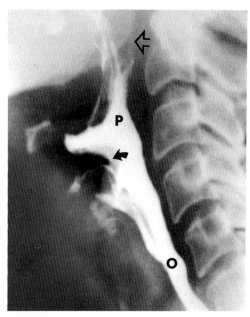

3.6
Lateral radiograph of neck during swallowing of barium.

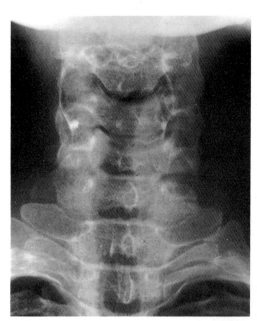

3.4
Frontal radiograph of neck.

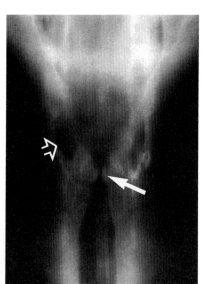

3.5
Planar tomogram of larynx.

Contrast media

It is often not possible to distinguish many organs by conventional radiography, especially in the abdomen. This problem can be overcome by the use of contrast media which are usually either pastes of inorganic barium salts for rectal or oral ingestion, or organic substances containing iodine for intravenous administration. To demonstrate the oesophagus and stomach, a sus-

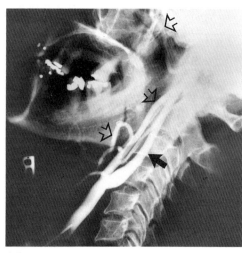

3.7
Angiogram of carotid arteries.

injection has been made into the common carotid artery and contrast medium is passing into the internal carotid artery (arrow) and branches of the external carotid artery (open arrows). This procedure, called **angiography**, can be applied to vessels almost anywhere in the body, including the chambers of the heart. A recent refinement has been the introduction of **digital subtraction angiography** which, by computer elimination of non-contrast-bearing surrounding tissue, provides a clear and well-defined image of the vessels.

Injection of contrast medium can also be made into a joint (**arthrogram**), into the bronchial tree (**bronchogram**), or into the spinal canal (**myelogram**). In **3.8** contrast medium has outlined the cervical spinal cord (C) and the cervical nerve roots (arrows) can be seen, running from the cord to the nerve root canals. In a **sialogram** injection is made into the duct of a salivary gland. In **3.9** the parotid duct has been catheterized and contrast medium can be seen filling the fine ducts of the parotid gland (arrows).

Ultrasound imaging

Sound waves travelling in medium are partly reflected when they hit another medium of different consistency. This produces an echo and the time taken for the echo to reach the source of the sound indicates the distance from the reflecting surface. Ultrasound imaging detects and analyses these echos.

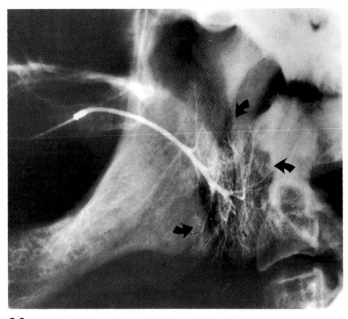

3.9
Sialogram of parotid salivary gland.

Ultrasound consists of high-frequency waves that cannot be detected by the human ear. When ultrasound travels in human tissue it undergoes partial reflection at tissue boundaries; thus a proportion of the sound waves return as an echo while the rest continue (**3.10**). Bone almost totally absorbs the sound and therefore no signal can be obtained from bone or the structures beyond. Neither can a signal be obtained from gas-containing viscera. These two facts are responsible for significant limitations to the use of this technique. However it has the great advantage that there are apparently no damaging effects at the sound energies used and it can therefore be used to monitor the developing fetus. In addition to its use in obstetrics, ultrasound is used to investigate the kidney, liver and biliary system, and thyroid gland. In **3.11** a normal thyroid lobe (T) is seen on the left, but the right lobe is occupied by a fluid-filled cyst (C) which transmits sound and therefore produces no echos. The absence of an image in the centre of the field examined is due to total

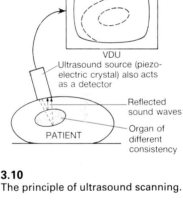

VDU
Ultrasound source (piezo-electric crystal) also acts as a detector

Reflected sound waves

PATIENT

Organ of different consistency

3.10
The principle of ultrasound scanning.

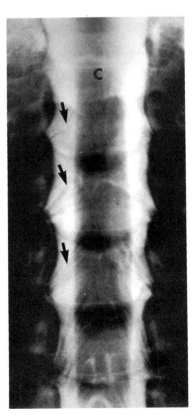

3.8
Myelogram of cervical spinal cord; A–P view.

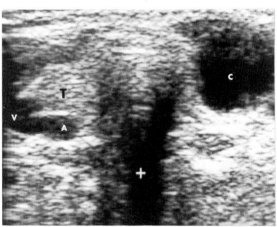

3.11
Ultrasound image of thyroid gland; A–P view.

reflection of the sound at the tissue/air interface of the anterior tracheal wall. Note the common carotid artery (A) and internal jugular vein (V).

Nuclear imaging

Nuclear medicine may be broadly defined as the use of radioactive isotopes ('radionuclides') in the diagnosis and treatment of disease. Interest to the student of anatomy arises from that branch of nuclear medicine called nuclear imaging or 'scintigraphy', whereby a 'map' of the uptake of radionuclide in a given organ system or pathological lesion can be produced by means of an instrument called a gamma camera. Analysis of such an image (which depends not only on morphology but also on function) is then of use in the diagnosis of the patient's problem.

The basis of the gamma camera is a large, flat crystal of sodium iodide which converts into light rays the gamma rays emitted from radionuclides which have been injected into the patient. These light rays strike a photosensitive surface and cause the emission of electrons. The latter are amplified by a photomultiplier and eventually electrical pulses are formed. These electrical pulses can be used to produce an image on a television screen or be used as input to a computer system (**3.12**).

A number of radionuclides are used in a form described as 'radiopharmaceuticals'. The principal radionuclide used is technetium-99m (which is very safe for the patient in terms of absorbed radiation dose, and has a convenient half-life of 6 hours), but others are also available, e.g. iodine-123, thallium-201, gallium-67, indium-111. These radionuclides are coupled to various compounds designed to deposit them in the organ of interest. For example, technetium-99m coupled to sulphur colloid will be phago-

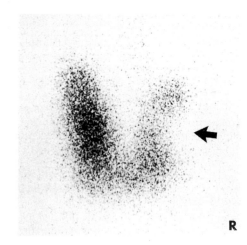

3.13
Nuclear image of thyroid gland; A–P view.

cytosed by macrophages and so be deposited in the reticuloendothelial system. Technetium-99m phosphates will deposit in bones. Iodine-123 as sodium iodide will be taken up by the thyroid. In **3.13** there is normal uptake in the left lobe, but in the middle of the right lobe uptake is absent, because the normal tissue is replaced by a lesion that does not accumulate the radionuclide (arrow). In this patient ultrasound of this area showed a thyroid cyst (**3.11**).

Cross-sectional imaging

Transmission computed tomography

Transmission computed tomography (TCT or CT) is similar to conventional radiography in that a beam of X-rays is passed through the body and is measured after emerging. The difference between CT and conventional X-ray methods is that a very narrow beam is used and an array of highly sensitive photoelectric cells is substituted for photographic film. The beam is rotated around the patient and density measurements are made from many different angles. The data is analysed by a computer and the whole image is displayed on a television monitor (**3.14**), dense structures such as bone conventionally being shown at the white end of the spectrum.

CT images are usually true axial sections, although coronal sections can be taken of the head by positioning the patient appropriately. However, computer programs are available whereby a sagittal or coronal image can be built up from data derived from successive axial images. CT produces extremely clear and finely-detailed cross-sectional radiographs without any superimposition of surrounding structures. Thus it is possible to see organs or small differences in density caused by disease which are almost impossible to demonstrate by conventional radiography. **3.15** is a CT (computed tomograph) of the brain showing the ventricles lying

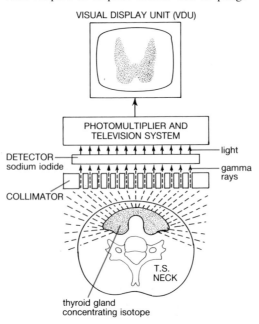

3.12.
The principle of nuclear imaging.

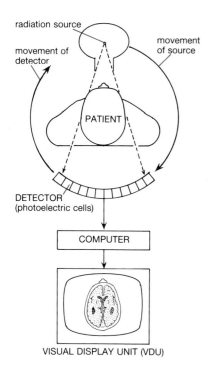

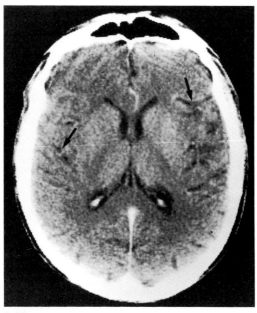

3.15
CT showing horizontal section of skull, frontal sinuses, and brain.

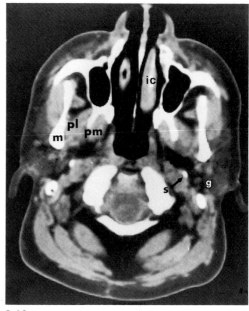

3.16
CT showing horizontal section through face and neck.

3.14
The principle of computed tomography.

centrally; grey matter lies peripherally, adjacent to the sulci on the surface of the brain (arrows). **3.16** is a CT taken just below the hard palate and shows the resolution of cross-sectional examinations. For example, the mandible (m), medial (pm) and lateral (pl) pterygoid muscles, styloid process (s), inferior nasal concha (ic), and parotid gland (g) can all clearly be seen.

Emission computer tomography (ECT)

This is based on the same principle as transmission computed tomography, except that it depends on gamma rays emitted from a radionuclide concentrated in an organ, rather than on a transmitted X-ray beam (**3.17**). Radionuclides, such as technetium-99m (and the others mentioned above), give off single photons (packets) of gamma rays so that, when these are used for emission tomography, the acronym becomes SPE(C)T or single-photon emission-(computed) tomography.

In **3.18**, technetium-99m-labelled amines have been taken up in the cerebral tissue of a young boy with epilepsy. The distribution of activity in the SPECT cross-section is normal except in the right frontal lobe where there is an area of reduced activity, indicating abnormal function and possibly the focus of disease.

There is also a variety of emission computed tomography which uses radionuclides known as positron emitters and so the technique is referred to as positron emission (computed) tomography (PET). The common radiopharmaceutical used for this is fluorine-18 deoxyglucose and its distribution reflects metabolism of glucose. Fluorine-18 gives out a positron (posi-

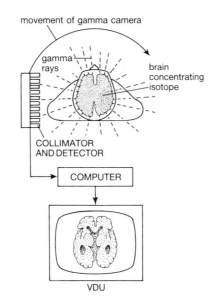

3.17
The principle of emission-computed tomography (ECT). In PET detectors surround the patient's head.

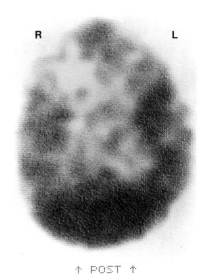

3.18
SPECT image of brain of patient with epilepsy.

tively charged electron) in one direction and an electron in the opposite direction. Neither particle can exist as such for any length of time and is soon annihilated by a particle of opposite charge. Energy is released during the encounter and as a result, two high-energy annihilation gamma rays emerge in opposite directions. It is only when these two rays are detected in coincidence that an 'event' is registered on the moni-

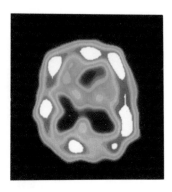

3.19
Normal PET scan (using fluorine-18-
deoxyglucose) through the brain.
The computer-generated colours
indicate the intensity of metabolism

tor tube. In this way cross-sectional 'metabolic maps' can be produced (**3.19**).

Magnetic resonance imaging (MRI)

MRI is a diagnostic technique based on radio signals emitted from resonating atoms within the body. This nuclear magnetic resonance effect depends upon the fact that atoms carry a charge and can therefore be regarded as small magnets. When a large magnetic field is applied across a body there is a tendency for the nuclei to line up with this field. Nuclei spin and, rather like spinning tops, can be displaced from their axis by a field applied at an angle to the main one. If the second field is of an appropriate frequency, the atoms can be made to resonate. Since the atomic nuclei carry a charge, any movement sets up radio signals which can be detected outside the body. The interest in MRI centres on the fact that the behaviour of the resonating nuclei depends upon the atoms surrounding them, in other words their chemical environment. Thus the radio signals measured outside the patient can reflect the biochemistry of the area under examination.

The MRI scanner looks rather like a CT machine in that the patient lies on a couch and enters a gantry which in this case is a very large electromagnet. Sectional magnetic resonance images of the area being examined are computed and these superficially resemble CT images. However, CT is based on the *density* of tissues to X-rays, whereas magnetic resonance images reflect some aspects of the *biochemical composition* of the tissues. As NMR is a purely electronic technique, it is also possible to obtain images in any plane. **3.20** is a midline sagittal MR image of the head. Beneath the cerebral hemispheres the brainstem (B) and cerebellum (C) can clearly be seen, surrounded by cerebrospinal fluid, which appears dark. The tongue (T) and sinus in the sphenoid bone (s) are also seen.

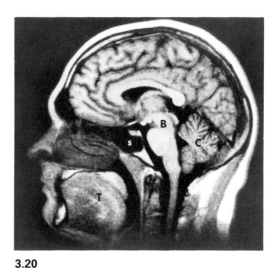

3.20
Magnetic resonance image showing sagittal section of head.

CHAPTER 4

Introduction to the head and neck

Man is characterized by an upright stance and by the possession of a brain which is larger and more convoluted on its outer surface (cerebral convolutions) than that of any other mammal. The evolution of the upright stance and those modifications required to support an increased vertical load are associated with inevitable changes in the structure of the head and neck, but whether such changes preceded, evolved in parallel with, or resulted from bipedalism, can only be surmised.

In Man the skull is balanced on the top of the flexible bony spine (**4.1**). This arrangement, which enables considerable mobility of the head and neck, is associated with several mechanisms which prevent instability. The articulation of the skull with the foramen magnum has shifted forward relative to its position in lower animals, thereby altering the centre of gravity of the heavy skull and enabling it to be balanced with greater ease on the uppermost cervical vertebrae. The spinous processes of the cervical vertebrae are united by a strong, somewhat elastic ligament which gives passive resistance to flexion. Short muscles anchor the base of the skull to the upper cervical vertebrae, while longer muscles connect the skull and cervical vertebrae to the lower spine, the upper ribs, and the shoulder girdle. These muscles provide postural stability for the head and neck and are responsible for the extensive range of movements which markedly increase the total effective field of Man's stereoscopic vision. The skull consists of a neurocranium, which surrounds and protects the brain and the inner and middle ears which are concerned with hearing and balance, and a viscerocranium which forms the bony skeleton of the face. The neurocranium has foramina which provide for the entry and exit of the spinal cord, of cranial nerves, and of the vessels which supply the brain. The viscerocranium provides the skeletal framework for the eyes, nose, mouth, larynx, and pharynx. The deep sockets for the eyeballs and their associated muscles face anteriorly for stereoscopic vision. The skeleton of the nose, which is flat in comparison with that of snouted non-primates, extends as a double passage from the external nares to the nasopharynx. Its roof is lined by olfactory epithelium sensitive to molecules which stimulate the sensation of smell, its walls are lined with a vascular epithelium which cleanses, warms, and moistens the inspired air, and it is connected to a series of air sinuses within the cranial bones. The upper (maxilla) and lower (mandible) jaws provide an anchor for the teeth and are moved by the muscles of mastication. In the floor of the mouth, both the mandible and hyoid bone anchor the base of the tongue and, with the upper jaw and base of the skull, give bony attachment to the pharynx which conveys food from the mouth to the oesophagus and air from the nose or mouth to the larynx.

Food entering the mouth must, if solid, be cut up and ground by the teeth, mixed with saliva, formed into a bolus of food by the muscular tongue and cheeks, and swallowed. The tough epithelium of the tongue is sensitive not only to common sensation (pain, temperature, touch), but also to taste which enables the nature of the oral contents to be analysed and, if necessary, rejected. The act of swallowing must move the bolus from the mouth to the pharynx and thence to the oesophagus, at the same time preventing its access to either the nasopharyngeal or laryngeal parts of the airway with which the pharynx is also continuous. At other times, the airway must be open for breathing.

Air is inhaled through the nose or mouth and passed via the pharynx into the larynx and thence to the trachea and lungs; it is exhaled through the same passages. The larynx has become specialized by the presence of a controllable pair of vocal folds. The vocal folds both enable the larynx to be closed off and intrathoracic pressure to be increased during coughing, lifting, and straining, and also can produce vibrations in the column of air being exhaled, producing sounds of varied pitch. One of the most important developments in Man, which sets him apart from other primates and lower mammals, is a superior power of communication. To the ability to gesture with the limbs and face has been added the ability to manipulate the larynx, tongue, palate, and face to produce speech and language which, combined with manual dexterity, led eventually to the written word. Speech and language are determined, not so much by the anatomy of the sound-producing tissues as by the sophistication of their nervous control.

Nervous control of the head and neck is regulated and controlled by cranial and spinal nerves which contain sensory, motor, and autonomic fibres. The cranial nerves supply the special senses (vision, hearing, taste, and smell); provide sensory innervation to the face, airway, and mouth, and the alimentary canal as far as the splenic flexure of the colon; and control the movements of the eyes, muscles within the ear, the jaws, tongue, larynx, pharynx, and upper alimentary tract. Autonomic (sympathetic and parasympathetic) fibres regulate the size of the iris and lens of the eye; the secretion of the

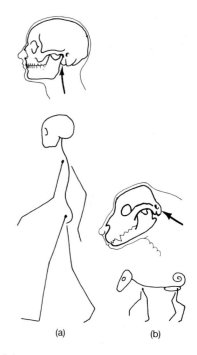

4.1
Differences in articulation between skull and spine in: (a) Man; (b) quadruped.

lacrimal and salivary glands, the mucous and serous glands of the airway and alimentary tract, and the specialized glands of the alimentary tract; they also regulate the diameter of most blood vessels in the head and neck but have little effect on those intracranial vessels which supply the brain. Cell bodies of parasympathetic fibres lie in the brainstem and their axons emerge with, and are distributed by, cranial nerves. Cell bodies of postganglionic sympathetic fibres, many of which are distributed to their targets along blood vessels, are situated in the cervical sympathetic chain, which derives its preganglionic control from the upper thoracic spinal cord.

The head and neck receive their blood supply from two major vessels on each side — the common carotid arteries which divide into external and internal main branches, and the subclavian artery. Those tissues lying outside the cranium are supplied largely from the external carotid artery, while the subclavian artery supplies the structures in the root of the neck. The brain and inner ear are supplied by both the internal carotid artery and the vertebral artery, a branch of the subclavian artery which ascends through the cervical vertebrae. Venous blood from within the cranium drains into a series of venous sinuses formed by the fibrous tissue lining the skull, and thence into the internal jugular vein. The scalp, face, and neck drain partly into the internal jugular vein, partly into the anterior and the external jugular veins, and partly into the subclavian vein, all of which eventually unite to form the superior vena cana. There are few valves in the veins of the head and neck, and venous hydrostatic pressure is minimized by the upright stance.

Lymphatic drainage is unnecessary within the central nervous system because of the special conditions that apply there, but the remainder of the head and neck is richly supplied with lymphatic tissue, nodes, and vessels. The face and scalp drain into groups of nodes arranged in collar fashion around the junction of neck and skull. Lymph from tissue lying deep to the mucous membrane of the mouth, alimentary tract, and airway drains into nodes which lie around these channels, and thence, with lymph from the face and scalp, into vertical lymphatic channels which lie along the main veins and along the course of which are numerous nodes. The lymph finally enters the venous system at the base of the neck on each side.

The head and neck contain a number of important endocrine glands. Beneath the brain lies the pituitary gland — its neural part (posterior pituitary) is formed by the axons of hypothalamic neurons; its anterior part consists of endocrine cells linked to the hypothalamus by portal veins which convey factors from the brain essential for its control. The thyroid gland lies around the anterolateral aspect of the larynx and trachea, and embedded in its posterior aspect are two pairs of parathyroid glands. Hormones from all these glands enter the venous system and thence the systemic circulation.

The overall arrangement of all these components follows a general plan. Posteriorly lies the central nervous system, encased in the skull and vertebrae and their associated ligaments and muscles. Immediately beneath the cranial cavity lie the eyes anteriorly and the ears laterally. In the midline lie the various structures of the airway and alimentary system, and in the angle between these central structures and the posterior musculoskeletal column lie the vessels and nerves that supply and control them. Various layers of condensed connective tissue divide up these compartments, and enable some to move easily on others, and surround and support the entire neck.

Any general introduction to a part of the body which attempts both to give a broad outline of the systems and to highlight features of particular interest inevitably leaves much unsaid. We hope that readers will make their own notes in the margins of the book.

In conclusion, make sure that you link your study of the gross anatomy with that of the function and microscopic anatomy of the tissues. Supplement the brief embryological sections of this book with a specialized text and, when studying the cranial and spinal nerves, make notes on their central nervous connections.

CHAPTER 5

Development of the head and neck

G. M. MORRISS-KAY

Ancestry and human development

Division of the body into head and trunk is a fundamental characteristic of vertebrates, the two components having different functions and different origins. This division of structure and function is thought to have evolved in relation to forward movement in an aquatic environment, perhaps originally in the larvae of prevertebrate chordate animals similar to modern sea-squirts (**5.1a**): the trunk generated forward propulsion while the head contained the brain and special sense organs, together with an oropharynx which served the dual functions of feeding and respiration. Water was sucked in through the mouth by depression of the floor of the oropharynx, then passed out through gill slits, thereby oxygenating the respiratory surface (gills). In all fishes, the gills have a blood supply from aortic arches, and are supported by the branchial (gill) skeleton, which is moved by muscles supplied by motor components of cranial nerves (**5.1b**). The human embryonic pharynx reflects this ancestry, except that perforated gill slits do not form.

One fundamental characteristic of the trunk is its segmental nature, which is based on embryonic somites and their derivatives (sclerotome, myotome, and dermatome). Somites are derived from paraxial mesoderm which lies on either side of the notochord. This pattern of development is clear in the neck and extends rostrally into the occipital region of the head. Rostral to the first (occipital) pair of somites, the cranial paraxial mesoderm is not obviously segmented, but some of its derivatives (e.g. the extrinsic eye muscles) are equivalent to somite-derived structures.

Taking both an evolutionary and a developmental perspective, then, the developing human head and neck may be considered as having three interrelated components: the **neurocranium**, which consists of the brain and the paired special sense organs; the **oropharynx**, which is associated with feeding, breathing, vocalization, and conduction of incoming sound; and other structures derived from **paraxial mesoderm** (**5.2b**). This pattern is reflected in the innervation of the head and neck (**5.3c**) which comprises:

Cranial nerves

● Nerves supplying the special sense organs:

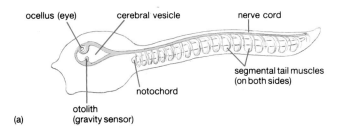

(a)

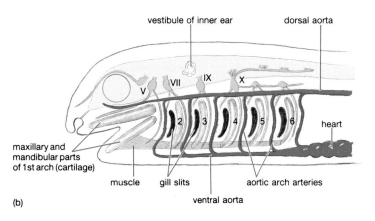

(b)

5.1
(a) Head and trunk specialization in sea-squirt larva.
(b) Organization of the head of an ancestral vertebrate. Caudal to each gill slit lies a cartilage (stippled). The cartilages of the first arch have become modified to form maxillary and mandibular components of the jaws.

olfactory (I), optic (II), and vestibulocochlear (VIII) nerves.

● Nerves supplying the embryonic pharynx: trigeminal (V), 1st arch; facial (VII), 2nd arch; glossopharyngeal (IX), 3rd arch; vagus (X), 4th and 6th arches; accessory (XI), 6th arch.

● Nerves supplying muscles derived from paraxial mesoderm: oculomotor (III), trochlear (IV), and abducent (VI) to extrinsic eye muscles; hypoglossal (XII) to muscles of the tongue derived from occipital somites.

Cervical nerves

● C1, C2, C3 innervate strap muscles associated with the embryonic pharynx.

● C1–C8 innervate muscles of the cervical spine.

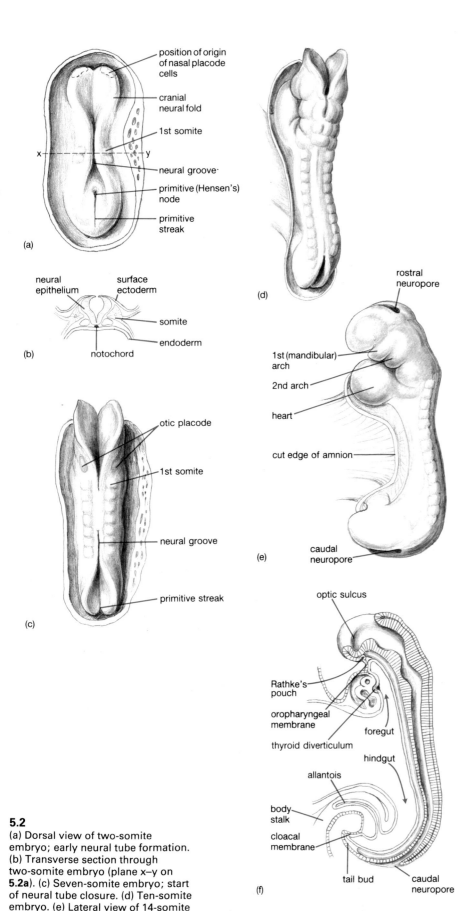

(a)

position of origin of nasal placode cells

cranial neural fold

1st somite

x————————y

neural groove

primitive (Hensen's) node

primitive streak

(b)

neural epithelium

surface ectoderm

somite

endoderm

notochord

(c)

otic placode

1st somite

neural groove

primitive streak

(d)

rostral neuropore

(e)

1st (mandibular) arch

2nd arch

heart

cut edge of amnion

caudal neuropore

(f)

optic sulcus

Rathke's pouch

oropharyngeal membrane

thyroid diverticulum

foregut

hindgut

allantois

body stalk

cloacal membrane

tail bud

caudal neuropore

5.2
(a) Dorsal view of two-somite embryo; early neural tube formation. (b) Transverse section through two-somite embryo (plane x–y on **5.2a**). (c) Seven-somite embryo; start of neural tube closure. (d) Ten-somite embryo. (e) Lateral view of 14-somite embryo. (f) Sagittal section through 14-somite embryo.

Organization of the embryonic head into neurocranial, oropharyngeal, and paraxial mesodermal components is clear at 4–5 weeks of development; therefore it is helpful to have a good understanding of embryonic structure at this stage, and to divide craniofacial development into this early morphogenetic phase and a subsequent phase during which the 4–5 week embryonic structure is transformed into the fetal form.

Early morphogenesis of the craniofacial region

During early stages of development the embryo undergoes major changes of shape which involve the folding of epithelial sheets and migratory movements of whole populations of mesenchymal cells.

The materials of early craniofacial development

Early embryos have only two types of tissue organization: epithelium and mesenchyme. **Epithelial cells** have a free apical surface, a basal surface with (or sometimes without) a basement membrane, and lateral surfaces in contact with each other by specialized junctions. **Mesenchymal cells** have no obvious polarity; they are separated by relatively large spaces which are filled with extracellular matrix. The molecular composition of the extracellular matrix around mesenchymal cells, and of the basement membrane of epithelial cells, has important influences on cell behaviour during morphogenesis. Changes in molecular composition, brought about by changes in gene expression, are correlated with the onset and termination of cell migration and with the generation of epithelial curvatures. Extracellular matrix molecules are linked by receptor molecules within the cell membrane to contractile elements of the cytoskeleton. Hence changes in composition of the matrix may initiate changes in cell shape or migratory behaviour. Cell surface adhesion molecules are also involved in these morphogenetic mechanisms.

Tissue components derived from all three **germ layers** are involved in early morphogenesis (**5.2**).

● **Ectoderm**: surface ectoderm gives rise to the epidermis and placode-derived structures; neural ectoderm forms the neural tube, neural crest, and olfactory epithelium.

● **Mesoderm**: primary mesenchyme, which is derived from the primitive streak, gives rise to segmented and unsegmented paraxial mesoderm and to lateral plate mesoderm. There is no intermediate mesoderm in the head.

● **Endoderm**: embryonic endoderm, which is also derived from the primitive streak, forms the epithelial lining of the pharyngeal gut and its diverticula.

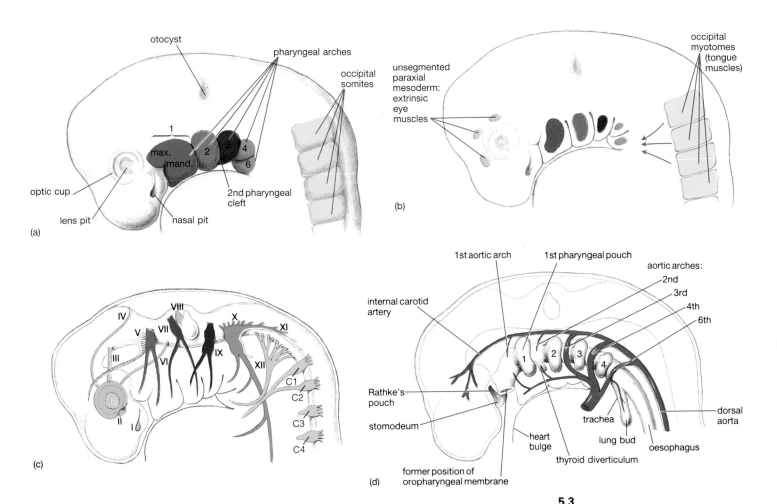

5.3
Views of 7-mm embryo to show: (a) early development of nose, eye, ear, pharyngeal arches, and somites; (b) paraxial mesoderm forming cranial striated musculature (see Table 1 for details of pharyngeal arch-derived muscles); (c) innervation of special sense organs, pharyngeal arch derivatives, and paraxial mesoderm derivatives by cranial and cervical spinal nerves; (d) arterial supply to the developing head and neck. Key 5.3c: blue, special senses; yellow, III, IV, VI, XII, and cervical; red, V$_2$ and V$_3$ (1st arch); red/white, V$_1$; orange, VII (2nd arch); purple, IX (3rd arch); green, X (4th and 6th arches); uncoloured, spinal XI.

Early development of the cranial nervous system

The nervous system of the head is derived from the **cranial neural tube**, **cranial neural crest**, and **placodes**.

● **Cranial neural tube (5.2)**. The cranial region is the first part of the neural plate to differentiate in the trilaminar embryo. It appears first at 18 days, before the formation of any somites. It is much broader than the neural plate of the spinal region; the caudal region of the hindbrain forms a transition between the two. Forebrain, midbrain, and hindbrain regions can be distinguished at an early stage. Closure of the neural tube starts in the upper cervical region, then proceeds in a rostro-caudal sequence in the trunk, and in a caudo-rostral sequence in the head. The cranial neural folds are at first convex in shape, becoming V-shaped and then concave in profile prior to closure. The rostral neuropore closes at 24 days, when the embryo has 20 pairs of somites. (The caudal neuropore closes at 26 days, when the embryo has 25 pairs of somites.)

After closure of the neural tube, accumulation of cerebrospinal fluid causes the brain vesicles to enlarge. A controlled pattern of growth and morphogenesis results in the formation of the cerebral hemispheres, cerebellum, and other components of the brain.

The pituitary gland has a dual origin: the neurohypophysis (posterior pituitary) forms as a downgrowth of the diencephalon (caudal part of the forebrain); the adenohypophysis (anterior, intermediate, and tuberal parts of the pituitary) is derived from a diverticulum (Rathke's pouch) from the roof of the oral cavity, positioned just rostral to the tip of the notochord and the oropharyngeal membrane (5.2f).

● **Cranial neural crest (5.4)**. Neural crest cells originate within the lateral region of the neural plate in all regions except the forebrain, in which crest cells arise from the optic region only. They cannot be distinguished morphologically from non-crest cells until they start to emigrate from the neural epithelium by conversion to mesenchyme. In the trunk and caudal hindbrain regions, neural crest cells begin to migrate only

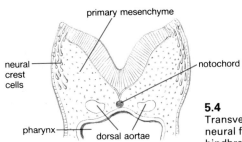

5.4
Transverse section through the open cranial neural folds of an eight-somite embryo (rostral hindbrain level) to show migration of neural crest.

after the neural tube has closed, but in the major part of the head of human and other mammalian embryos the migration begins at an earlier stage.

The cells migrate along defined pathways to specific destinations, where they differentiate into a variety of cell types. All cranial nerve ganglion cells (i.e. the afferent components of all cranial nerves except I and II) are derived from neural crest, together with a smaller contribution from placodes (see below). Cranial neural crest cells also migrate out of the head to form the postganglionic parasympathetic neurons innervated by the vagus nerve.

Non-neural derivatives of the cranial neural crest include connective tissue, cartilage, bone, and melanocytes (see the section on 'Cranial mesenchyme').

● **Placodes**. Placodes are localized thickenings of surface ectoderm. There are two categories of placodes:

(a) Those which contribute to formation of the paired special sense organs.
(b) Those which contribute sensory neurons to the cranial nerves.

The latter (epibranchial placodes) are situated at the proximal end of each pharyngeal arch, overlying the developing cranial sensory ganglia; cells detach from them to contribute sensory neurons to the ganglia of the pharyngeal arch nerves (V, VII, IX, and X). The otic placode falls into both categories, forming both the inner ear and part of the ganglion of the vestibulocochlear nerve (VIII).

Formation of neurons and glia

Neuroblasts (developing neurons) form from the neuroepithelium of the neural tube, from the neural crest, and from placodes. Within the neural tube neuroblasts may migrate from their original positions; neural crest- and placode-derived neuroblasts have already undergone migration. In their final sites neuroblasts form axons and dendrites by the extension of processes (neurites), the cell body of the differentiated neuron remaining at the end-point of migration.

The extending processes make synaptic contacts with other developing neurons.

The **sensory** (afferent) components of cranial nerves (except I and II) are formed by the extension of neurites from each developing cranial nerve ganglion into the neural tube and into the structure it innervates. The **motor** (efferent) components are formed by neurites extending peripherally from neuroblasts within the neural tube.

Neuroepithelial and neural crest cells (but not placode-derived cells) also form **neuroglia**, the non-neuronal cells of the nervous system.

Early development of nose, eye, and inner ear

The lining of the nasal cavity originates as a group of neuroepithelial cells at the rostral edge of the forebrain region of the neural plate, in a position analogous to that of the neural crest in all other regions (**5.2a**). This group of cells slides into the adjacent surface ectoderm before neural tube closure to form **nasal placodes** on the front of the face. These become concave as microfilaments close to the apical (outer) surface of the placodal cells contract, forming the nasal pits. The olfactory epithelium forms from the deepest region of the nasal pits, where they are closest to the developing forebrain. The olfactory nerve (I) forms by growth of neurites from the olfactory epithelium into the olfactory region of the forebrain (see **6.3.1**, **6.3.2**).

The retina and optic nerve (II) form as an outgrowth from the forebrain. They appear first as an optic sulcus in each of the open neural folds (**5.2f**); from this develops the **optic cup** (**5.3a**). The lens forms from the **lens placode**, a thickening of surface ectoderm which is induced by a tissue interaction with the underlying optic cup. The lens placode forms a lens pit as described for the nasal placode, except that the pit continues changing in shape to form a closed lens vesicle beneath the surface ectoderm (see **6.8.1**).

The inner ear originates within the surface ectoderm as the **otic placode**, which is induced by a tissue interaction with the adjacent hindbrain neural folds. The otic placode becomes concave, then separates from the surface ectoderm to form the otic vesicle (**5.6**). The otic vesicle develops further to form the vestibule and cochlea (see **6.9.1**). Some cells detach from it to form sensory neurons of the vestibulocochlear ganglion (VIII). Neurites from this ganglion extend into the developing otic vesicle and into the hindbrain, to form the vestibulocochlear nerve (VIII).

Cranial mesenchyme (5.7)

During and after the period of neural tube closure the embryonic head and neck contain a large amount of mesenchyme which is derived from three sources: **neural crest, paraxial mesoderm**, and **lateral plate mesoderm**. Mesenchymal cells differentiate into a variety of cell types. The contribution of cells from each of

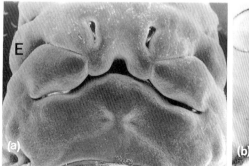
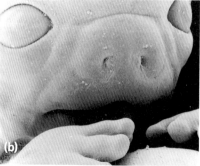

5.5
Development of face (a) 6-week embryo, E eye; (b) 8-week embryo, after fusion of the facial swellings (see **6.3.1** for details).

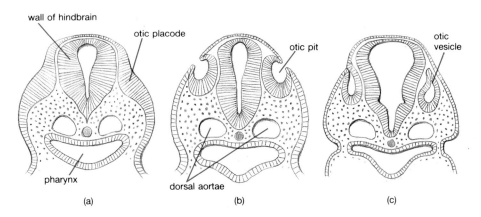

wall of hindbrain
otic placode
otic pit
otic vesicle
pharynx
dorsal aortae
(a)   (b)   (c)

5.6
Development of otocyst (a) 20 days (nine somites); (b) 23 days (16 somites); (c) 30 days (30 somites).

the three sources to connective tissues, including skeletal tissues, is indicated in **5.7**. Some of the information that follows is derived from transplantation studies in avian embryos; there is good evidence that it is also valid for mammalian embryos, including Man.

● **Neural crest**. Cranial neural crest cells form a greater variety of cell types than those of the trunk. Like trunk crest cells they form neurons and neuroglia; they also migrate to the skin where they form melanocytes. Only cranial neural crest cells have the potential to differentiate into chondrocytes, osteocytes, and connective tissues including dermis, tendons, ligaments, and the walls of blood vessels (excluding the endothelium). Through such differentiation they make a major contribution to the development of the face and ear, and some important craniofacial abnormalities are associated with abnormal neural crest cell migration. Cranial neural crest cells provide the mesenchymal components of the salivary, thyroid, parathyroid, and thymus glands, and type 1 cells of the carotid body. They also make important contributions to the developing heart, in particular the mesenchyme of the truncus arteriosus and conus cordis, and aortic arch arteries (Vol. 2, Chapter 5, Seminar 4), which may explain why congenital abnormalities of the face are sometimes associated with congenital heart defects.

● **Paraxial mesoderm**. All the striated muscle of the head and neck is derived from paraxial mesoderm. Around the developing eye it forms the extraocular muscles. In the occipital region the paraxial mesoderm segments to form four pairs of somites, which behave similarly to those in the trunk (Vol. 1, Chapter 8, Seminar 1): the sclerotome spreads around the caudal part of the hindbrain to form much of the occipital bone (see **6.1.1**); the myotome migrates ventrally and rostrally to provide the intrinsic muscles of the tongue, taking with it its segmental nerve supply (hypoglossal nerve, XII); the dermatome contributes to the dermis and connective tissue of the occipital region. Cervical somites are typical

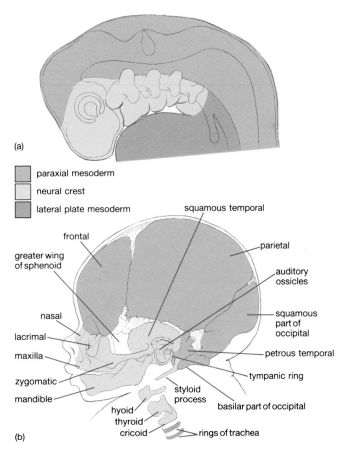

(a)

■ paraxial mesoderm
□ neural crest
■ lateral plate mesoderm

squamous temporal
frontal
greater wing of sphenoid
parietal
auditory ossicles
nasal
squamous part of occipital
lacrimal
maxilla
petrous temporal
zygomatic
tympanic ring
mandible
styloid process
hyoid
basilar part of occipital
thyroid
cricoid
rings of trachea
(b)

5.7
Contributions of neural crest, paraxial and lateral plate mesoderm to (a) the connective and skeletal tissue of the developing head in a 7-mm embryo and (b) the skull and anterior neck skeleton at term.

trunk somites whose sclerotome forms vertebrae, myotome forms striated muscle, and dermatome contributes to the dermis and connective tissues of the neck.

● **Lateral plate mesoderm**. The lateral plate is, as its name suggests, the most lateral mesoderm in the embryo. In the early three-layered embryo the lateral plate of the two sides extends to the rostral pole of the embryo. This region, which contains the developing heart, moves ventrally and caudally during cranial neurulation so that there are no derivatives of lateral plate mesoderm in the head, although it does contribute some connective tissues in the anterior part of the neck (dermis, tracheal cartilages, and possibly the lower part of the larynx).

The embryonic pharynx (5.3, 5.8)

The embryonic pharynx has five pairs of pharyngeal arches (also called branchial or visceral arches), which are numbered 1, 2, 3, 4, and 6 by analogy with non-mammalian vertebrate embryos, the 5th arch having been lost in mammals (**5.3a**). Each arch consists of a layer of mesenchyme sandwiched between two epithelia: pharyngeal endoderm and surface ectoderm. Most of this mesenchyme is derived from neural crest cells, but there are some primitive streak-

derived cells (paraxial mesoderm) which give rise to the pharyngeal musculature. The depressions between each arch are called pharyngeal clefts on the outer surface and pharyngeal pouches on the inner surface (**5.3d**). Muscle, cartilage, connective tissue, and an aortic arch artery form from the mesenchyme of each arch (**5.3b,d**) (Table 1). The cranial nerve which supplies motor fibres to arches 1 to 3 also has a branch to the arch in front; these 'pretrematic' (i.e. in front of the cleft) branches do not supply muscle (**5.3c**).

Pharynx-derived structures include the face, palate, tongue, thyroid gland, pharynx, larynx, and the external and middle parts of the ears.

Transformation of the four- to five-week embryo to the fetal form

The following brief description summarizes some of the changes which transform the embryonic to the fetal craniofacial structure. All of these processes are described in more detail in the introductory section of the relevant seminars.

Skull (5.7)

The skull is formed from local cranial mesenchyme. Thus, the neurocranium, which surrounds the brain and paired sense organs, is mainly formed from paraxial mesoderm while the viscerocranium, which is derived from the embryonic pharynx and forms the face and ear regions, is derived from neural crest (**5.7a**). Both neurocranium and viscerocranium have components formed by endochondral and intramembranous ossification.

Face and secondary palate

The face and palatal shelves form as a series of prominences composed of surface ectoderm and underlying neural crest-derived mesenchyme. These undergo shifts in position and adjacent swellings fuse (**5.5**). Failure of fusion leads to clefts of the face, lip, and palate. During this phase of development, changes in proportion occur so that the eyes move to the front of the face, while growth in length of the face lowers the tongue from its position between the palatal shelves, allowing them to move together and fuse in the midline (see **6.3.1**, **6.4.1**, **6.4.2**).

Pharynx (5.8 and Table 1)

The first pharyngeal cleft forms the external auditory meatus, but the others are covered over leaving no trace in the fetus, except rarely when the cervical sinus persists as a cervical cyst. Auricular hillocks form as small swellings on either side of the first cleft; these combine to form the pinna of the ear.

The first pharyngeal pouch, together with part of the second, forms the middle ear cavity and

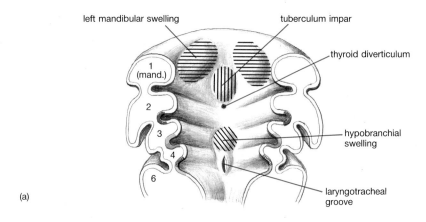

(a)

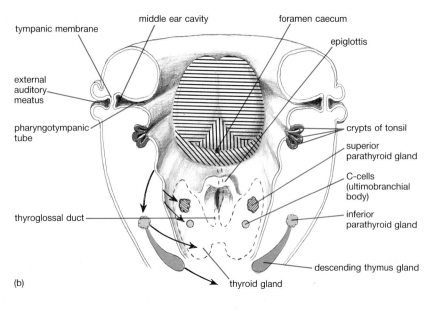

(b)

5.8
Diagrams of the floor of the developing pharynx to show contributions of the various pharyngeal arches, clefts, and pouches to the formation of the tongue, larynx, ear, tonsil, parathyroid, thymus, and thyroid glands.

pharyngotympanic tube; it extends around the developing auditory ossicles so that these are bathed in amniotic fluid until birth. The endoderm of pharyngeal pouches 2–4, together with the adjacent mesenchyme, forms the palatine tonsils, parathyroid and thymus glands, and the calcitonin-secreting cells of the thyroid gland. The body of the thyroid forms from endoderm and mesenchyme of the distal tip of the thyroglossal duct; the duct descends as a midline diverticulum from the floor of the first arch (**5.8b**). The tongue forms from three swellings on the floor of the first arch and one on the third arch. The fourth arch makes a small contribution to the back of the tongue but the second makes little or no contribution. This pattern of development is reflected in the innervation of the tongue (p. 59). The third, fourth, and sixth arches form the fetal pharynx and larynx.

Table 1. Derivatives of the pharyngeal arches and their nerve supply

Arch	Nerve	pre-trematic branch (br) / post-trematic branch	Muscles (ms)	Bones, cartilages, and ligaments (ligt)	Artery
1 maxillary	V	maxillary division	—	incus	1st arch artery (transitory)
1 mandibular		mandibular division	→ ms of mastication, ant. belly digastric, mylohyoid, tensor palati, tensor tympani	malleus Meckel's cartilage, ant. ligt malleus sphenomandibular ligt	
		chorda tympani / greater petrosal nerve	—		
2 hyoid	VII	facial	→ stapedius, stylohyoid, post. belly digastric, ms of facial expression, auricle & scalp, buccinator	stapes, styloid process, stylohyoid ligament, hyoid — lesser horn and upper part of body	stapedial artery (transitory)
		tympanic br IX	—		
3	IX	glossopharyngeal	→ stylopharyngeus	hyoid — greater horn and lower part of body	common carotid artery, 1st part internal carotid artery
		pharyngeal br X	→ sup. & middle pharyngeal constrictors, palatal ms except tensor palati		
4		sup. laryngeal br X	→ inf. pharyngeal constrictor cricothyroid	thyroid cricoid arytenoid } laryngeal cartilages	arch of aorta, stem of right subclavian artery
6	Cranial XI / X	recurrent laryngeal br X	→ laryngeal ms except cricothyroid		pulmonary arch ductus arteriosus

Seminar 1

Skull and cervical skull

The **aim** of this seminar is to study general features of the **skull** and **cervical spine**. The skull comprises a number of bones, which can be grouped into (*a*) the **neurocranium** which protects the brain, the eyes, and the ears and (*b*) the **viscerocranium** which forms the skeletal framework of the face, nose, air sinuses, and mouth. Parts associated with the eye, nose, and mouth will be considered again with those regions. The cervical spine consists of the atlas (C1), axis (C2), and C3–C7 vertebrae (see Vol. 1, Chapter 8). This seminar is longer than others in this volume and may take more than the usual 2–3 hours study time.

Development of the skull
(6.1.1, 6.1.2)

The skull develops from embryonic mesenchymal tissue which surrounds the developing brain and from the first and second pharyngeal arches. In general, the **neurocranium** is formed from paraxial mesoderm while the **viscerocranium** is derived from neural crest tissue.

The **neurocranium** is initially formed by either membranous or cartilaginous models of separate bones which later become ossified (**6.1.2**). The skull base begins development as separate cartilaginous elements which develop around the cranial end of the notochord (**6.1.1a**). The pattern of their growth is affected by the presence of cranial nerves and blood vessels (**6.1.1b**). The three occipital sclerotomes

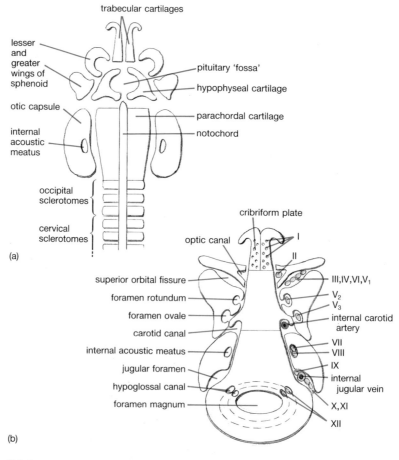

6.1.1
Development of base of skull: (a) components of cartilaginous neurocranium; and (b) nerves and vessels which pass between them.

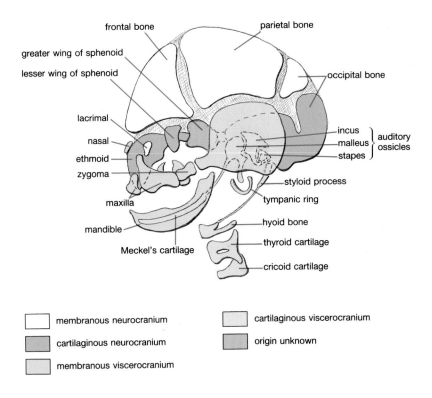

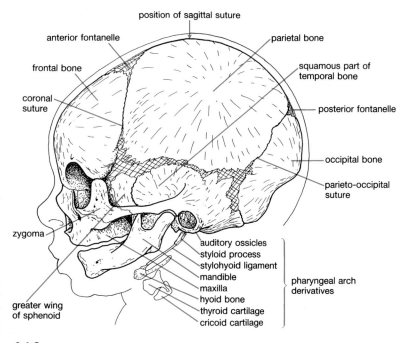

6.1.2
Skeletal components of skull and anterior part of the neck. Cervical vertebrae (not shown) are formed from cervical sclerotomes.

fuse with the cartilaginous plate around the notochord to form the base of the occipital bone, and extend around the hindbrain and foramen magnum just as sclerotomes in the trunk form vertebrae, so that the foramina through which the hypoglossal nerves (XII) leave the skull are equivalent to intervertebral foramina. Other cartilages, which eventually ossify to form the ethmoid bones, grow around the nasal sacs while a similar encapsulation occurs around the developing inner and middle ears to form the petrous temporal bone.

The vault of the neurocranium is formed by ossification of membranous models of the frontal bones, parietal bones, squamous parts of the temporal bones, and superior part of the occipital bone. At birth, fibrous connections, immature **sutures**, exist between these thin bones of the vault of the skull. At the junctions between (a) the frontal and parietal bones and (b) the parietal and occipital bones, the fibrous areas between the bones are relatively large and are termed respectively, the **anterior fontanelle** and the **posterior fontanelle (6.1.3)**. This arrangement enables the cranial vault to mould during parturition, but no such deformity of the bones at the base of the skull can occur during birth. As the head of the fetus starts to descend the birth canal, its anterior fontanelle can usually be felt by a finger inserted into the anal canal of the mother, thus enabling the orientation of the fetal head to be ascertained. After birth, any skull deformation which has occurred should return to normal in 2–3 days.

6.1.3
Lateral aspect of skull at birth.

The membrane bones of the skull grow both at their margins with the sutures and also by appositional growth on their outer aspect, while the inner aspect is eroded to contain the

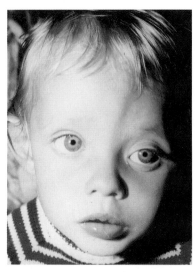

6.1.4
Facial asymmetry due to premature fusion of sutures on left side of skull.

6.1.5
Achondroplasia.

growing brain. If the sutures of the cranial vault fuse prematurely (craniosynostosis; **6.1.4**), abnormalities of skull growth and therefore shape will occur which, if uncorrected, increase intracranial pressure and can damage the brain.

Abnormal development of endochondral bone (due to defective formation of type II collagen) leads to a relatively common form of dwarfism — achondroplasia (**6.1.5**). Those affected have short limbs and pear-shaped heads; the bones of the skull vault which develop in membrane are normally formed, but those at the base which develop in cartilage remain small.

The **viscerocranium**, like the neurocranium, is formed by ossification of both cartilaginous and membranous elements. The cartilaginous elements are formed from neural crest-derived mesenchyme of the pharyngeal arches. The first arch consists of mandibular and maxillary components but only the mandibular contains a cartilage (Meckel's cartilage). Meckel's cartilage becomes replaced by membrane bone which forms lateral to it and partially invests it to form the major part of the mandible, though secondary endochondral bone contributes later to its head and coronoid process. The position of the cartilage is indicated, in the adult, by the mandibular canal. If Meckel's cartilage fails to

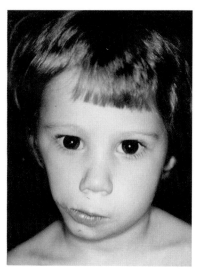

6.1.6
Facial asymmetry due to maldevelopment of right mandibular arch.

develop properly on one side, a marked facial asymmetry results (**6.1.6**). Intramembranous ossification within mesenchyme of the maxillary prominences and the area between them forms the maxilla, zygomatic bone, and

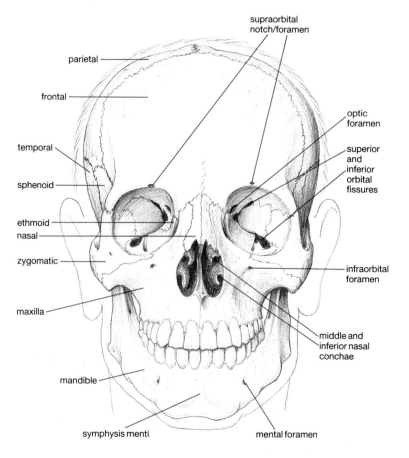

parietal
frontal
temporal
sphenoid
ethmoid
nasal
zygomatic
maxilla
mandible
symphysis menti
supraorbital notch/foramen
optic foramen
superior and inferior orbital fissures
infraorbital foramen
middle and inferior nasal conchae
mental foramen

6.1.7
Anterior aspect of skull.

squamous part of the temporal bone (part of the neurocranium).

The dorsal end of the 1st arch cartilage ossifies to form two auditory ossicles, the malleus and incus. Ossification of the dorsal end of the 2nd arch cartilage (Reichert's cartilage) forms a third auditory ossicle (stapes) and the styloid process, while its ventral end forms the lesser horn and upper part of the body of the hyoid bone (see **6.1.2**).

For development and ossification of vertebrae see Vol. 1, Chapter 8.

A. Examination of the skull

Before studying an isolated skull, look at an articulated skeleton and note that, in bipedal Man, the axis of the skull (as defined by oral and nasal cavities) is situated almost at a right angle to the upper cervical vertebrae (see **4.1**).

During your examination of the skull you should determine which bony features can be felt in the living person, by palpation of yourself and your partner. Structures passing through the foramina of the skull will be given only in the section on internal features.

Neurocranium — external features

Anterior aspect (6.1.7)
Take a skull with the skull cap in position and examine its anterior aspect. The anterior aspect of the neurocranium consists of the **frontal bone** which also forms the roof of the orbit. Feel the **supraorbital ridges** and note the position of the **supraorbital notch (or foramen)** about half-way along the ridge. On the lateral side of the orbit the frontal bone meets the frontal process of the zygomatic bone and, on the medial side, it meets the nasal bone at the bridge of the nose. Trace the extent of the frontal bone and its sutures with the parietal, sphenoid (greater wings), and zygomatic bones.

Dorsal aspect (6.1.8a,b)
Examine the dorsal aspect of the skull cap. Identify the **coronal suture** between the frontal bone and parietal bones, the midline **sagittal suture** between the two parietal bones, and the lambdoid suture between the parietal bones and the occipital bone. The sagittal suture joins the coronal suture at the **bregma** and the lambdoid suture at the **lambda**. Small isolated parts of cranial vault bones within the sutures are called **Wormian bones**. Occasionally, development of the right and left halves of the frontal bone leaves a midline (metopic) suture running forward from the bregma.

Posterior aspect (6.1.9)
The posterior aspect of the skull consists largely of the **occipital bone** with its **foramen magnum** through which the brainstem passes to become continuous with the spinal cord. On either side of the foramen magnum, identify the **occipital condyles** which articulate with facets on the lateral masses of the atlas. Posterior to the condyles you may find an **emissary foramen** from which veins draining the diploë (marrow

cavity) of the skull emerge. In the midline is an easily palpable midline prominence, the **external occipital protuberance**, and, extending laterally on either side of the external occipital protuberance, are **superior nuchal lines**. Between the superior nuchal lines and the foramen magnum are further bony lines and markings associated with the attachment of extensor spinal musculature.

Lateral aspect (6.1.10)
The lateral aspect of the neurocranium is formed from the **frontal** and **parietal bone** superiorly, and the **squamous part of the temporal bone** and the **'greater wing' of the sphenoid bone** inferiorly. The area where the frontal, parietal, temporal, and sphenoid bones

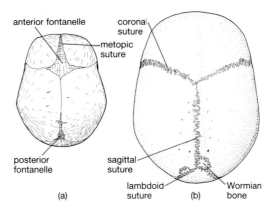

6.1.8
Dorsal aspect of skull: (a) newborn; (b) adult.

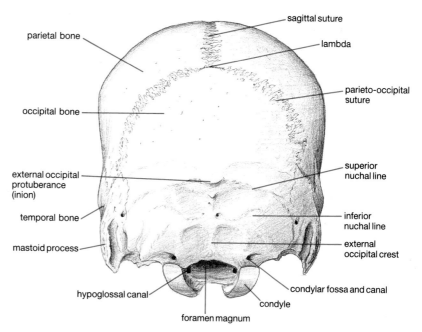

6.1.9
Posterior aspect of skull.

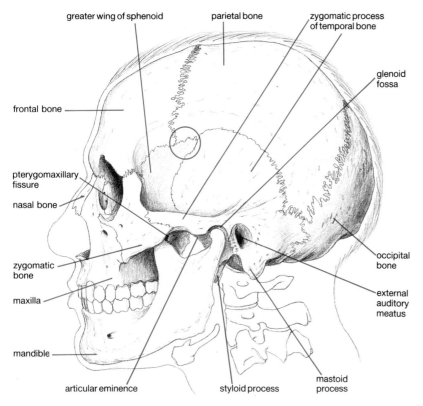

6.1.10
Lateral aspect of skull, pterion circled.

meet is called the **pterion**, an area of potential weakness (p. 107).

On the **temporal bone** locate the canal of the outer ear (**external acoustic meatus**), the **tympanic plate** which forms its anterior and inferior margin, and the **mastoid process**, a marked bony protuberance which extends downward behind the external ear. It may be relatively solid but more often it is honeycombed by mastoid air cells which connect with the middle ear. Note the **zygomatic process** of the temporal bone, a strut which projects forward from the squamous part of the temporal bone to articulate with the **zygomatic** (cheek) bone to form the **zygomatic arch**. The prominence on the under-surface of the zygomatic process of the temporal bone is the **articular eminence** and behind this lies the **mandibular (glenoid) fossa** with which the head of the mandible articulates via an articular disc.

Basal aspect (6.1.11)

The base of the neurocranium consists essentially of two parts. The anterior part is formed from the **frontal** and **ethmoid** bones which roof, respectively, the orbits and nose. The posterior part consists of a thick, midline platform of bone comprising the **body of the sphenoid, basisphenoid**, and **basiocciput**, and lateral parts formed from the **greater wings of the sphenoid**, the **petrous**, and inferior **squamous** parts of the temporal bone. Behind these again lie the foramen magnum and occiput.

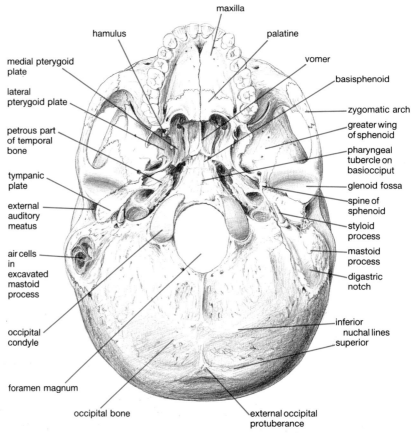

6.1.11
Basal aspect of skull.

Look first into the orbit (see p. 00) and examine its roof, the **orbital plate of the frontal bone**, and, posteriorly, the **lesser wing of the sphenoid** perforated by the round **optic canal**. Between the orbits lies the **ethmoid bone**. Remove the skull cap, hold the skull against the light, and look up into the nasal cavities. You will see that, on either side of the midline, the roof of the nose is formed by a perforated bridge of bone, the **cribriform plate** of the ethmoid. This unites two roughly rectangular masses of bone (**labyrinths**) which lie on either side of the upper part of the nasal cavity and form much of the medial walls of the orbit. The labyrinths are honeycombed with air cells which communicate with and drain through several openings into the nose. Passing downward into the nose from the midline of the cribriform plate is the **perpendicular plate** of the ethmoid which forms part of the midline nasal septum (see p. 99).

Posteriorly, the base of the cranium is formed in the midline by the body of the sphenoid (largely hidden from view by the skeleton of the nose) and the conjoined basisphenoid and basiocciput on which is found the midline pharyngeal tubercle. On each side, a horizontal extension from the body of the sphenoid forms its **greater wing**, and medial and lateral **pterygoid plates** project downward; between them is a space, the **infratemporal fossa**. The greater wing of the sphenoid thus has an infratemporal and a temporal surface; between the two is a ridge, the infratemporal crest. The posterior extremity of the greater wing of the sphenoid bears a small process, the **spine of the sphenoid**. Posterior to the greater wing of the sphenoid and lateral to the central bony platform locate parts of the temporal bone: medially, the **petrous part** (rock-like, it degenerates last after burial) of the temporal bone is separated from the central mass by the foramen lacerum; more laterally, the **styloid process** (a rapier-like projection passing downward), the **mastoid process**, the **tympanic plate**, and the **mandibular (glenoid) fossa**. Between the tympanic plate and the mandibular fossa a small part of the petrous temporal bone obtrudes.

The sphenoid and temporal bones are pierced by foramina which transmit structures between the neck and the cranial cavity. The presence of the foramina and numerous sutures creates a potential weakness in the base of the skull which can therefore be fractured by severe trauma. At the anterior margin of the greater wing of the sphenoid is the **inferior orbital fissure** and the greater wing is pierced by the **foramen ovale** and the **foramen spinosum**. Anterior to the petrous part of the temporal bone is the bony part of the **auditory (pharyngotympanic) tube**. The large **carotid canal** penetrates the petrous temporal bone from below. Insert a flexible probe into the canal and note its orientation and tortuous course. The **stylomastoid foramen** lies, as its name suggests, between the styloid and mastoid processes. Between the petrous temporal bone and occiput lies the **jugular**

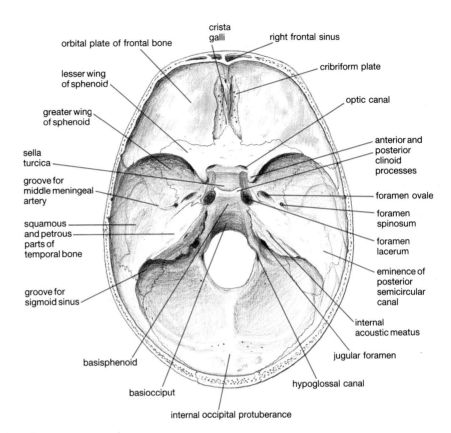

crista galli
orbital plate of frontal bone
right frontal sinus
lesser wing of sphenoid
cribriform plate
greater wing of sphenoid
optic canal
sella turcica
anterior and posterior clinoid processes
groove for middle meningeal artery
foramen ovale
squamous and petrous parts of temporal bone
foramen spinosum
foramen lacerum
eminence of posterior semicircular canal
groove for sigmoid sinus
internal acoustic meatus
jugular foramen
basisphenoid
hypoglossal canal
basiocciput
internal occipital protuberance

6.1.12
Interior of skull.

foramen, and the occiput itself is pierced by the large **foramen magnum** and, anterior to each articular condyle, by the **hypoglossal canal**. Posterior to the condyles are variable posterior condylar canals.

Neurocranium — internal features (6.1.12)

Remove the skull cap and examine the **cranial cavity** within the skull. You will see that there are three major concave areas — the anterior, middle, and posterior cranial fossae (**6.1.13**). Structures which pass through the various foramina and fissures are shown in **6.1.12** and noted in the text in square brackets [ ].

Anterior cranial fossa

The **anterior cranial fossa** supports the frontal lobes of the brain; it consists of the orbital plates of the frontal bone between which lies the perforated **cribriform plate** of the ethmoid [olfactory nerve fibres from the nose] from which projects the upright **crista galli** (cockscomb) to which the falx cerebri is attached (p. 78). Posterior to each orbital plate lies the 'lesser wing' of the sphenoid, from which the **anterior clinoid process** projects posteriorly.

Middle cranial fossa

The **middle cranial fossa** supports the temporal lobes of the brain, the hypothalamus, and pituitary. Its floor is formed centrally by the

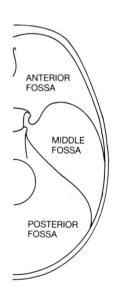

ANTERIOR FOSSA

MIDDLE FOSSA

POSTERIOR FOSSA

6.1.13
Cranial fossae.

body of the sphenoid and laterally by the greater wing of the sphenoid and anterior surface of the petrous temporal bone; its lateral wall is formed from the extension of the greater wing of the sphenoid and from the squamous part of the temporal bone.

The body of the sphenoid contains the sphenoid air sinuses. Dorsally, this bone is thin and shaped like a Turkish saddle (**sella turcica**) which supports the pituitary gland and is therefore known as the **pituitary (hypophysial) fossa**. The anterior clinoid processes lie lateral to the pituitary fossa and to the S-shaped sulcus for the internal carotid artery on either side of the body of the sphenoid. Posteriorly, the fossa is limited by a rectangle of bone, the dorsum sellae, from which project the **posterior clinoid processes**. Laterally, the body of the sphenoid is continuous with its greater wings which, in turn, articulate with the frontal, parietal, and temporal bones. By holding the skull up to the light, note how thin the bone is at this junction.

Locate the foramina of the middle cranial fossa. Anterolateral to the pituitary fossa, the **optic canal** [optic nerve, ophthalmic artery, and sympathetic plexus] passes to the orbit between the body and lesser wings of the sphenoid. At the anterior extremity of the middle cranial fossa the **superior orbital fissure** [oculomotor (III), trochlear (IV), ophthalmic division of trigeminal (V), and abducent (VI) cranial nerves and the ophthalmic veins] communicates with the orbit between the lesser and the greater wings of the sphenoid before they fuse laterally. Pass a probe through the forward-facing **foramen rotundum** [maxillary division of V]; it passes from the cranial cavity into the **pterygomaxillary fissure** between the maxilla and pterygoid plates of the sphenoid bone. Posterolateral to the foramen rotundum locate the **foramen ovale** [mandibular division of V] and **foramen spinosum** [middle meningeal artery]; pass a probe through each of them into the infratemporal fossa and, from the foramen spinosum, trace vascular markings made by the middle meningeal vessels laterally toward the pterion. At the medial extremity of the petrous temporal bone is the foramen lacerum. Pass a flexible probe through the carotid canal from below; it will appear in the upper part of the foramen lacerum; from here trace the impression of the internal carotid artery forward on the lateral wall of the body of the sphenoid. On the anterior face of the petrous temporal bone locate two small canals passing anteromedially toward the foramen lacerum and foramen ovale [greater and lesser superficial petrosal (autonomic) nerves]; locate also the elevation produced by the superior semicircular canal of the inner ear.

Posterior cranial fossa

The **posterior cranial fossa** supports the cerebellar hemispheres and, centrally, the brainstem. Laterally, it is formed by the petrous and squamous parts of the temporal bones and the squamous part of the occipital bone; centrally,

by the conjoined basiocciput and basisphenoid (**clivus**).

On the posterior surface of the petrous temporal bone identify the **internal acoustic meatus** [facial (VII), vestibulocochlear (VIII) cranial nerves; labyrinthine artery] which leads from the cranial cavity to the inner and middle ear, within the petrous temporal bone. Locate the **jugular foramen** between the petrous temporal and occipital bones [glossopharyngeal (IX), vagus (X), accessory (XI) cranial nerves; inferior petrosal and sigmoid venous sinus] and the **hypoglossal canal** [hypoglossal (XII) cranial nerve].

The interior of the skull is marked by venous sinuses. Pick up a skull cap (**6.1.14**); identify the midline sagittal suture which unites the parietal bones and, on its inner aspect, a linear depression which marks the position of the **superior sagittal sinus**. Alongside this depression you may find rounded pits marking the position of arachnoid granulations which return CSF to the venous system (see **6.7.4**). Follow the marking for the superior sagittal sinus backward; opposite the external occipital protuberance it will diverge to one side (usually the right) to become continuous with a groove for the **transverse sinus** which, at the junction of the occipital and petrous temporal bones, curves downward to the jugular foramen as the **sigmoid sinus**. Markings of the transverse and sigmoid sinuses are found on the opposite side also (see p. 79). At the suture between the apex of the petrous temporal bone and the basisphenoid/occiput lies the **inferior petrosal sinus**, while the **superior petrosal sinus** makes a small impression on the upper posterior border of the petrous temporal bone.

A severe blow to the skull may cause the bone of the skull to fracture. If the blow is caused by a sharp object then a localized, depressed fracture occurs. By contrast, if the blow is caused by a rounded or blunt object or follows impact of the head against a flat surface, fractures may extend over large areas of the skull. Such a fracture may damage underlying tissues: vessels

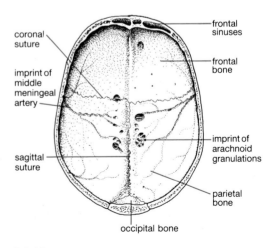

coronal suture

imprint of middle meningeal artery

sagittal suture

frontal sinuses

frontal bone

imprint of arachnoid granulations

parietal bone

occipital bone

6.1.14
Interior of skull vault.

(venous sinuses, veins, arteries); cranial nerves as they pass through the various foramina; the paranasal sinuses; and the brain within its dural sac (see **6.1.26**). If the force of the blow causes the brain to move within its protective cerebro-spinal fluid and strike the opposite side of the cranial cavity, injury to the brain may occur opposite the part of the skull that has been struck (contre-coup injury).

Viscerocranium (facial bones)

The bones of the face (**6.1.7**) are slung beneath the anterior part of the cranium. They comprise the paired **nasal** and **maxillary** bones and the **zygomatic** bones which provide strong lateral buttresses between the facial skeleton and the cranium; also the ethmoid, lacrimal, palatine, and vomer.

The two **nasal bones** join to form the bony bridge of the nose and can easily be distin-guished by palpation from the elastic cartilage that supports the antero-inferior part of the nose.

On either side, the **maxilla** forms the main part of the face and upper jaw. The maxillae unite in the midline beneath the nose to form the anterior part of its floor, while their **alveolar margins** carry the upper teeth (four incisors, two canines, four premolars, and six molars); in the infant the first 'milk' teeth lack premolars and the third molars ('wisdom teeth') (p. 61). Within the body of the maxilla is the pyramidal **maxillary air sinus**, the apex of which is directed laterally towards the zygomatic bone. Medially, at the base of the pyramid, the body of the maxilla forms the lateral wall of the nose through which the maxillary sinus drains (p. 50). Superiorly, the body of the maxilla forms most of the floor of the orbit. It sends a buttress-like process upward and medially to join the nasal and frontal bones, behind which it articulates with the lacrimal and ethmoid bones in the me-dial wall of the orbit. Laterally, the apex of the maxilla forms another buttress which articulates with the zygomatic bone.

Between the maxilla and the **lacrimal bone** is a depression which houses the lacrimal sac which drains via the nasolacrimal duct into the nose. In the floor of the orbit, the lateral margin of the maxilla is separated from the greater wing of the sphenoid by the **inferior orbital fissure**, from which a groove extends forward to the **infraorbital foramen** [infraorbital branch of maxillary V], which lies a finger's breadth lateral to the nostril.

Identify the horizontal **palatine processes** of the maxilla which meet in the midline to form the anterior part of the hard palate, which sepa-rates the anterior part of the nasal cavities from the mouth. Immediately behind the incisor teeth lies the **incisive foramen** [branches of maxillary V]. Posteriorly, the palatine processes of the maxillae articulate with the horizontal plates of the palatine bones.

Each **palatine bone** is L-shaped; in addition to a **horizontal plate** it has a **perpendicular plate** which forms part of the lateral wall of the

nose and continues into a process which forms a minute part of the floor of the orbit.

Examine the bony nasal septum (**6.1.15**); it is formed by the perpendicular plate of the ethmoid and small upward projections of the palatine processes of the maxilla and, between them posteriorly, a separate bone, the **vomer**, which also articulates with the body of the sphe-noid at the base of the skull. Within the nasal cavity three scroll-like extensions project inward from each lateral wall to form the skeleton of the conchae; the inferior of these is a separate bone.

Immediately posterior to the perpendicular plate of the palatine bone lie two further per-pendicular sheets of bone, the **medial and lateral pterygoid plates** (**6.1.16**). These are downward extensions of the sphenoid bone which articulate with the maxilla via the palatine bone, thus forming another buttress between the cranium and the facial skeleton. Note the **hamulus**, a small hook-like inferior projection of the medial pterygoid plate at the posterior extremity of the hard palate.

On your face and the skull, examine the front of the maxilla and its articulation with the

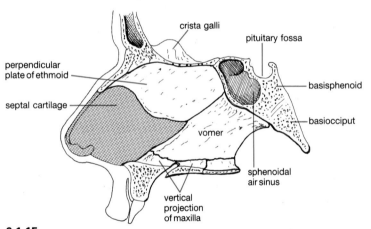

6.1.15
Nasal septum.

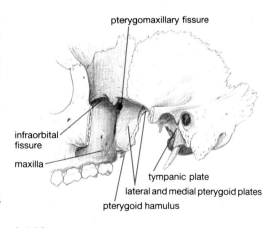

6.1.16
Pterygoid plates of sphenoid bone and pterygomaxillary fissure.

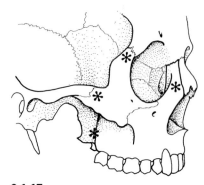

6.1.17
Buttresses (∗) between the face and cranium.

zygomatic bone which forms the skeleton of the cheek. The zygomatic bone has processes that project upward along the lateral aspect of the orbit to meet the frontal bone, and backward to meet the zygomatic process of the temporal bone. Its orbital surface is pierced by a foramen [zygomatic branch of maxillary V] which connects to foramina on its facial [zygomaticofacial nerve] and temporal [zygomaticotemporal nerve] aspects.

Severe blows to the face tend to fracture the facial skeleton at its point of buttress with the cranium (**6.1.17**):

● At the medial aspect of the orbit.

● At the articulations of the zygomatic with the frontal and temporal bones.

● At the articulation where the pterygoid plates meet the palatine and maxillary bones.

If the fracture is complete on both sides, then the whole facial skeleton may be pushed backward, thereby obstructing the upper respiratory tract.

Lower jaw (6.1.18; 6.1.19)

The **mandible** carries a set of teeth complementary to that of the maxilla and articulates with the cranium at the **temporomandibular joint**. Identify the **body** of the mandible; it lies horizontally and is continuous across the midline at the 'symphysis' menti (NB: This is not a joint of the secondary cartilaginous type (p. 5)). Posteriorly, at the **angle of the jaw**, the **ramus** of the mandible is set almost vertically and ends in the **condylar** and **coronoid** processes. The condylar process comprises a **neck** and a transversely-orientated articular **head**. The shape of the bone alters considerably with age (**6.1.20**).

On the inner aspect of the bone locate the **mandibular foramen**; its opening is protected by a bony projection, the **lingula**. The foramen leads into the **mandibular canal** which carries the inferior alveolar nerve and its corresponding vessels to the lower teeth and gums. The canal opens on to the outer surface of the mandible via the **mental foramen** through which some sensory nerves emerge. On the inner surface of the body of the mandible are markings of the attachment of muscles of the tongue and floor of the mouth. In the midline locate the **genial tubercles** and **digastric pits**; and, more laterally, a pronounced ridge, the **mylohyoid line**. Above and below the mylohyoid line are smooth fossae (∗) in which the sublingual and submandibular salivary glands are situated.

Hyoid bone (6.1.21)

The **hyoid** bone is U-shaped and lies horizontally above the thyroid cartilage in the neck. It has a midline body from which project, almost horizontally backward, two **greater horns** (cornua). Two much smaller **lesser horns** project upward at the junction of the body and greater horns. The hyoid is suspended from the styloid process

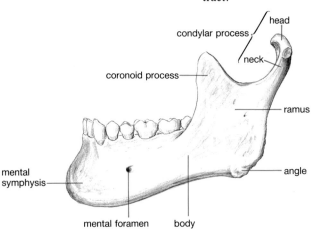

6.1.18
Lateral aspect of mandible.

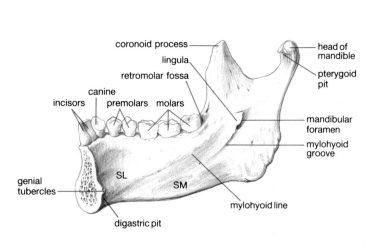

6.1.19
Medial aspect of mandible. Impressions of sublingual (SL) and submandibular (SM) salivary glands.

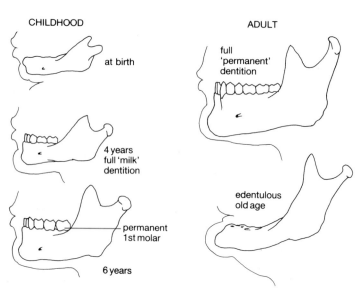

6.1.20
Age-related changes in shape of mandible.

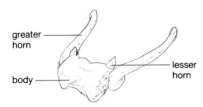

greater horn

lesser horn

body

6.1.21
Hyoid bone.

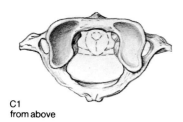

C1
from above

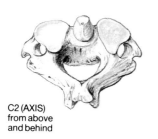

C2 (AXIS)
from above
and behind

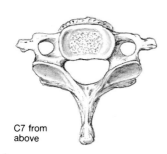

C7 from
above

6.1.22
Atlas (C1), axis (C2), and C7 vertebrae.

of the skull by the stylohyoid ligament, and forms an important anchorage for the tongue. Feel the body of the hyoid in the midline and its greater horns by holding the neck between thumb and index finger, between the mandible and larynx.

Cervical vertebrae (6.1.22)

For a fuller account of the spinal column see Vol. 1, p. 108; only specific features of the cervical spine will be covered here.

In the adult, the cervical spine has a gentle lordosis. This first becomes apparent at the time when the infant starts to raise its head. The entire cervical spine is relatively mobile on the relatively more fixed thoracic spine.

Each vertebra normally comprises a **centrum** or **body**; a **vertebral arch** comprising, on each side, a **pedicle** attached to the centrum and a broad **lamina**; a **spinous process**; two **transverse processes**; and **superior** and **inferior articular facets** which project from the pedicles so that each vertebral arch articulates with that of the vertebra above and below.

The first and second cervical vertebrae are extensively modified to permit the movements of the skull on the vertebral column.

● **C1 — The atlas**. The atlas, named after the Greek god who bore the world on his shoulders, lacks a centrum; during evolutionary development this became the odontoid process of the axis. The atlas comprises an **anterior arch**, with an **anterior tubercle** on its ventral surface to which the anterior longitudinal ligament is attached, and a facet on its dorsal surface for articulation with the anterior surface of the odontoid process. The larger **posterior arch** is grooved on its upper surface by the vertebral artery; its **posterior tubercle** represents the spinous process. Each **lateral mass**, which can be felt between the mastoid process and angle of the jaw, is pierced by a **foramen transversarium** for the vertebral artery, and extends laterally as the **transverse process**; the transverse ligament of the atlas (which retains the odontoid peg) is attached to its inner surface. Large concave **superior facets** articulate with the occipital condyles, and more flattened **inferior facets** articulate with the axis.

● **C2 — The axis**. The axis is easily distinguished by its **odontoid peg (dens)** which bears articular facets for the anterior arch of the atlas on its anterior surface, and for the transverse ligament of the atlas on its posterior surface, and facets for the attachment of the alar ligament on either side. The body of the axis has prominent flattened oval facets which articulate with the atlas. The **laminae** are very thick and the **spinous process** is large and bifid as in most other cervical vertebrae.

● **Cervical vertebrae C3–C6** are 'typical' cervical vertebrae. Each has a relatively small **body**, and a **transverse process** pierced by a foramen transversarium, bearing an **anterior tubercle** (costal element) and a **posterior tubercle**. The lateral margins of the upper surface of the body are slightly lipped, and the articular facets on the vertebral arches are flattened and orientated in a forward-sloping transverse plane. The **spinous processes** are bifid for the attachment of the nuchal ligament.

● **C7 — The 'vertebra prominens'** — is the first cervical vertebra that can readily be palpated as the examining hand descends the neck. Its spinous process is horizontal and not bifid, and the vertebral veins but not the artery pass through its foramen transversarium (which is often narrow).

At the lower end of the neck, the manubrium of the sternum, the clavicle, the first two ribs, the acromion and spine of the scapula, and the thoracic spinal column all provide bony attachments for the muscles and fascia of the neck. These are dealt with in Vols 1 and 2; see also pp. 131–2.

B. Radiology

Skull

Plain radiographs of the skull have an important place in clinical assessment despite the availability of more advanced forms of imaging, and the student should be familiar with their most important features. The four standard views are the lateral, posteroanterior (P–A), anteroposterior (A–P), and basal views. In order to displace the very dense petrous temporal bones from superimposition with the orbits, the beams used in P–A and A–P views are angulated. Remember that the part of the skull which shows most clearly is that which is positioned closest to the film. A number of other more specialized and local projections are also available, such as those for showing the paranasal sinuses or internal acoustic meatus.

Compare the P–A radiograph **6.1.23** with **6.1.7** and a skull. Identify as many named features as possible, in particular the frontal bone and supraorbital ridge, the zygomatic and maxillary bones, the petrous temporal bone and mastoid process, the nasal septum and inferior concha, and the mandible.

Similarly, compare the lateral radiograph **6.1.24** with **6.1.10**, identifying in particular the frontal, parietal, and occipital bones, the fronto-parietal and parieto-occipital sutures, the temporal bone, mastoid process and external auditory meatus, the maxilla and mandible.

Compare the basal radiograph **6.1.25** (and see **6.10.14**) with **6.1.11** (which does not show the mandible or cervical vertebrae) and identify the mastoid process and its air cells, the petrous temporal bone, the foramen magnum over which is superimposed the anterior arch of the atlas (arrows) and odontoid peg (O), the clivus (basiocciput and basisphenoid) and sphenoidal air sinus anterior to it, the vomer and septum of the nose, the margins of the orbit, and the mandible.

In addition to fractures (**6.1.26**), one of the most important abnormalities to look for in a plain radiograph of the skull of an adult is thinning or erosion of the dorsum sellae. This is caused by raised intracranial pressure which is,

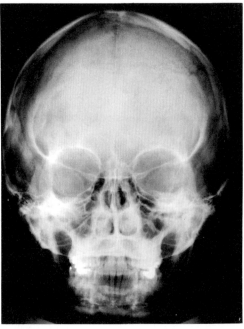

6.1.23
P–A radiograph of skull.

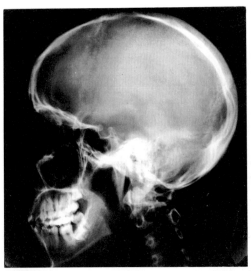

6.1.24
Lateral radiograph of skull.

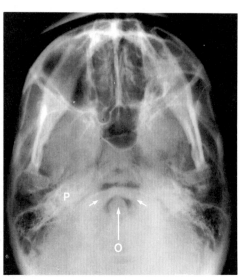

6.1.25
Basal radiograph of skull and upper cervical spine.

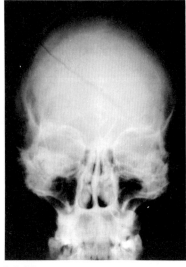

6.1.26
Fracture of skull cap.

in turn, often due to a rapidly growing intra-cerebral tumour exerting pressure on its sur-roundings. Most of the interior of the skull can withstand the increased pressure, but not the dorsum sellae (**6.1.27**; compare with **6.1.24**). In a child, increased pressure may splay the sutures (**6.1.28**). A tumour may calcify (**6.1.27**) and thus become visible on a plain film. Tumours may also cause **local** as opposed to generalized changes: a pituitary tumour may enlarge the pituitary fossa (see **6.7.8**; a tumour on the vesti-bulocochlear nerve ('acoustic neuroma') may ex-pand the internal auditory meatus. A tumour of the meninges (meningioma) can erode the adja-cent inner table of the skull. Displacement of a calcified pineal gland (which is normally in the midline) toward the opposite side can be caused by a tumour or a collection of blood within the skull, usually deep to the dura (subdural haema-toma, p. 78). However, not many pineal glands are calcified and then generally only in the elderly.

CT and MR images of particular parts of the skull are presented with specific regions in later seminars.

Cervical spine

Examine the A–P and lateral radiographs of the cervical vertebrae (**6.1.29**). Note the mild lordo-sis of the cervical spine in the upright position. Identify the individual vertebrae and their particular features: the anterior and posterior

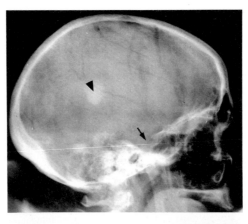

6.1.27
Erosion of posterior clinoid process (arrow) by raised intracranial pressure caused by rapidly growing meningioma which is partly calcified (arrowhead).

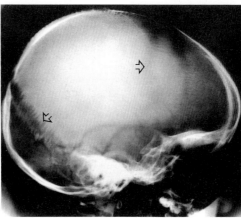

6.1.28
Radiograph of skull showing splayed sutures (arrows).

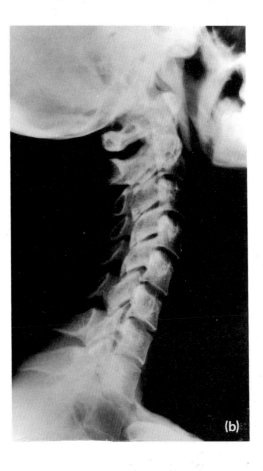

6.1.29
Cervical vertebrae: (a) A–P; (b) lateral.

arches of the atlas and its prominent transverse processes; the odontoid peg of the axis (**6.1.30**); the bodies, spinous processes, and transverse processes of C2–C7 and the upward-facing plane articular facets on their neural arches.

Identify the body and greater horn of the hyoid bone; parts of the laryngeal cartilages (p. 72) may become calcified and visible on radiographs.

Requirements:

Articulated skeleton; separate skull with removable skull cap and articulated mandible; hyoid bone.

Anteroposterior, posteroanterior, lateral, and basal radiographs of the skull.

Anteroposterior and lateral radiographs of the cervical spine.

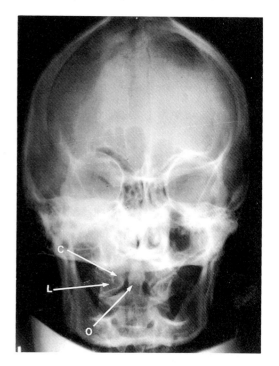

6.1.30
A–P view of atlas (lateral mass L) and axis (odontoid process O) viewed through the open mouth of an edentulous subject. Occipital condyle. (C).

Seminar 2

Movements and muscles of head, neck, and mandible

The **aim** of this seminar is to study the movements of the head and neck and of the mandible. The joints and muscles on which these movements depend should be studied in the living and by use of prosections. Muscles of the face, mouth, pharynx and larynx, eye, and ear will be considered separately in the seminars on these regions.

A. Living anatomy

Movement of the cervical spine

Examine the range of movements of the head and neck that you and your partner can achieve. The head and neck naturally move together, but the head can be flexed on the vertebral column until the chin touches the neck. Flex and extend your head on your spine, at the same time palpating the transverse processes of the atlas, the tips of which can be felt midway between the mastoid processes and angles of the jaw, to demonstrate that much of the flexion and extension movement occurs at the atlanto-occipital joint. Next, determine the degree of flexion and extension of the cervical spine that can be achieved while keeping the thoracic spine immobile (see **6.2.18**).

Qu. 2A *What limits extension of the head and neck?*

Rotate your head, at the same time keeping a finger on your laryngeal prominence (Adam's apple). This moves little as the head rotates because virtually all the rotation occurs at the atlanto-axial joint and not between other cervical vertebrae.

Move your head from side to side; this lateral flexion occurs at the atlanto-occipital joints and,

to a lesser extent, between the other cervical vertebrae (C3–C7). Now, push forward (protract) and then retract your head against resistance and note the contraction of prominent sternomastoid muscles (**6.2.1**) on the side of the neck during protraction. Rotate your head against resistance and demonstrate that the sternomastoid contracts on the opposite side to that toward which the face is rotated.

Finally, push your head to either side against resistance, and feel the contraction of the trapezius muscle (**6.2.2**) posterior to sternomastoid on the side of the neck (this muscle is more commonly tested by 'shrugging' the shoulders).

Movements of the lower jaw

Place an index finger immediately in front of the lobe of each ear, pressing against the neck of the mandible. Talk and chew gently, then open your mouth widely; close and open your mouth again and note the considerable forward movement of the head of the mandible beneath the zygomatic process of the temporal bone as the mouth is opened widely. Opening the mouth, therefore, consists of both depression of the mandible and protraction of its head; closing the mouth necessitates the reverse. During opening and closing the mouth, the position of the lingula of the mandible remains almost unchanged; in this way, the inferior alveolar nerve (which runs

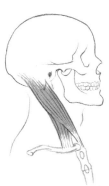

6.2.1
Sternomastoid.

6.2.2
Trapezius.

through the body of the mandible) is not stretched as the mouth is opened. In chewing movements, a variable amount of side-to-side movement of the mandible also occurs.

Place the fingers of both hands on either cheek and clench your jaw. You will feel masseter (see **6.2.15**) contract; define its anterior and posterior borders. Now place your fingers an inch or so above the zygomatic arch and again clench your teeth; here you will feel temporalis (see **6.3.16**) contract, but the contraction is more difficult to feel than that of masseter because a tough layer of deep (temporal) fascia overlies temporalis.

Get your partner to bare his teeth in an exaggerated grin and examine the position of the upper and lower incisor teeth. In the occluded (closed) position the incisor teeth of the lower

jaw are normally behind and in contact with those of the upper jaw.

B. Prosections

Atlanto-occipital, atlanto-axial, and cervical intervertebral joints

Articulate a skull and atlas vertebra together and examine the movements that can occur at the **atlanto-occipital joints** and which, together, form a single (ellipsoid) synovial articulation. On a prosection identify these joints at which the skull can flex and extend on the neck, thereby producing a nodding movement; a little lateral flexion is also possible.

Next, articulate an atlas and an axis together. Note that the prominent odontoid process (dens) articulates against the inner surface of the anterior arch of the atlas. On a prosection (**6.2.3**, **6.2.4**) note that the dens is held firmly against the atlas by the strong **transverse ligament of the atlas** which is attached on either side to the inner surface of the lateral mass of the atlas. The dens is thereby normally prevented from moving posteriorly and damaging the spinal cord which lies behind it. The transverse ligament is strengthened by vertical bands which attach to the occipital bone superiorly and the body of the axis inferiorly, thereby forming a **cruciate ligament**. Synovial joints exist between the dens and the transverse ligament and between the dens and the anterior arch of the atlas. Rotation of the head on the neck involves pivoting of the dens in this fibro-osseous ring and gliding movements between the superior facets of the axis and the inferior facets of the atlas.

Examine the roughened apex of the dens; the weak **apical ligament** (which contains a remnant of the notochord and sometimes a tiny bone, the proatlas) passes up to the anterior lip of the foramen magnum anterior to the cruciate ligament. Locate the strong **alar ligaments** which run laterally from smooth facets on the 'shoulders' of the dens to their attachments on the rim of the foramen magnum; these 'check' ligaments limit rotation between the atlas and axis, and lateral flexion of the atlanto-occipital joint; they also help prevent posterior movement of the dens.

In severe traumatic flexion injury to the neck, as in hanging, the transverse ligament (and alar ligaments) can rupture, allowing the dens to move posteriorly and damage the uppermost part of the spinal cord. If the dens fractures (typically at its attachment to the centrum of the axis), the skull and atlas can move forward on the axis thus shearing the spinal cord (**6.2.5**). Both injuries are usually rapidly fatal.

Identify the **anterior longitudinal ligament**, which runs anterior to vertebral bodies throughout the length of the vertebral column; it continues upward over the anterior arch of the atlas to the base of the skull as a narrow band, lateral to which the **anterior atlanto-occipital mem-**

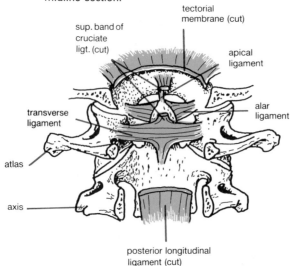

6.2.3
Ligaments of occipito-atlanto-axial region in midline section.

6.2.4
Ligaments of occipito-atlanto-axial region; posterior view with laminae removed.

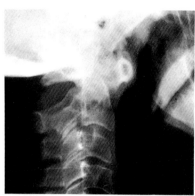

6.2.5
Dislocation of cervical spine between C1 and C2.

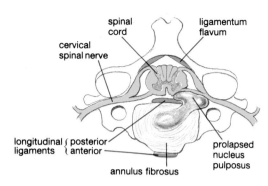

6.2.6
Prolapse of cervical disc.

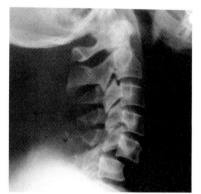

6.2.7
See **Qu. 2B**

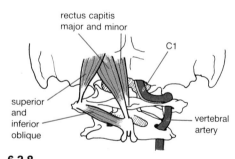

6.2.8
Suboccipital muscles.

brane blends with the capsules of the atlanto-occipital joints. Locate the upward extension of the posterior longitudinal ligament which is attached to the posterior aspect of vertebral bodies; it cannot attach to the atlas, but runs from the body of the axis to the anterior margin of the foramen magnum, lying posterior to the dens and cruciate ligament, and is known as the **tectorial membrane**. Between the posterior arch of the atlas and the foramen magnum is the **posterior atlanto-occipital membrane**, the equivalent of the ligamenta flava at lower levels.

Examine the joints between the lower cervical vertebrae and between C7 and T1. The intervertebral discs consist of an outer **annulus fibrosus**, the fibrous bands of which overlap and interdigitate with one another, and an inner proteoglycan-rich core, the **nucleus pulposus**, which lies nearer the posterior than the anterior surface of the disc. The cervical lordosis is produced by the anterior aspect of the disc being thicker than the posterior. As in the lumbar region, the nucleus pulposus of cervical discs can herniate either posteriorly toward the anterior part of the spinal cord, or anterolaterally toward the roots of the cervical spinal nerves, and compress these structures (**6.2.6**).

The angulated superior and inferior articular facets are so orientated as to permit flexion/extension and some lateral flexion. Each is surrounded by an articular capsule. The **anterior** and **posterior longitudinal ligaments** lie, respectively, anterior and posterior to the bodies of cervical vertebrae; and elastic **ligamenta flava** lie between the laminae of the vertebrae. In the neck, the **nuchal ligament** takes the place of the supraspinous and interspinous ligaments of lower vertebral levels. This large, fibroelastic membrane has a free margin which extends between the external occipital protuberance and the spinous process of C7, the vertebra prominens. Deeply, it is attached to the spines of cervical vertebrae, to the posterior tubercle of the atlas, and to the midline of the occipital bone. In bipedal Man it is much less elastic than in quadrupeds, but provides attachment for muscles of the back of the neck.

Qu. 2B *Comment on radiograph* **6.2.7**.

Muscles which move the cranium and cervical spine (6.2.8–6.2.10)

On prosections, study the muscles which move the atlanto-occipital, atlanto-axial, and other cervical intervertebral joints. They comprise four functional groups: flexors, extensors, lateral flexors, and rotators. Some of these are long muscles with good mechanical advantage; others are much shorter and have more of a postural role.

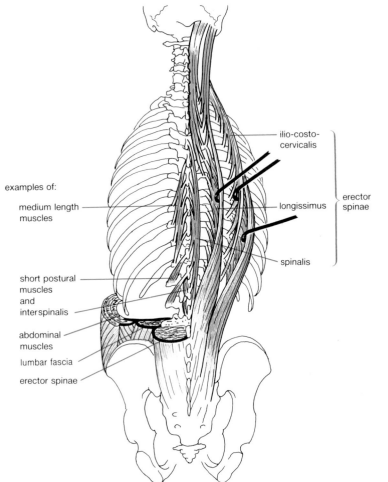

examples of:

medium length muscles

short postural muscles and interspinalis

abdominal muscles

lumbar fascia

erector spinae

ilio-costo-cervicalis

longissimus

spinalis

erector spinae

6.2.9
Long extensors of the cervical spine and skull.

Between the atlas and occiput locate *short straight* (rectus capitis anterior, lateralis, and posterior) muscles which pass from the anterior and posterior arches of the atlas to the occipital bone; these act in flexion and extension of the atlanto-occipital joint.

Short oblique muscles are attached either between the spine of the axis and the transverse process of the atlas (inferior oblique), or from the transverse process of the atlas to the base of the skull posteriorly (superior oblique); they act in rotation of the atlanto-axial joint.

Now identify the *long flexor and extensor* muscles of the spine. Anteriorly, longus colli is a long flexor attached to the anterior aspect of the vertebral column from atlas to T3; an upward extension, longus capitis, attaches between the basiocciput and the anterior tubercles of transverse processes of 'typical' (C3–6) cervical vertebrae. Posteriorly, a complex group of extensor muscles are attached to the occiput between the foramen magnum and superior nuchal line, to the nuchal ligament, and to posterior tubercles of the transverse processes of cervical vertebrae, as well as to lower levels of the vertebral column.

The *lateral flexor* muscles are the scalene group. On a prosection in which sternomastoid has been reflected (**6.2.10**) locate **scalenus anterior**. It is attached above to the anterior tubercles of the transverse processes of the typical (3–6) cervical vertebrae (anterior to cervical nerve roots) and passes downward and

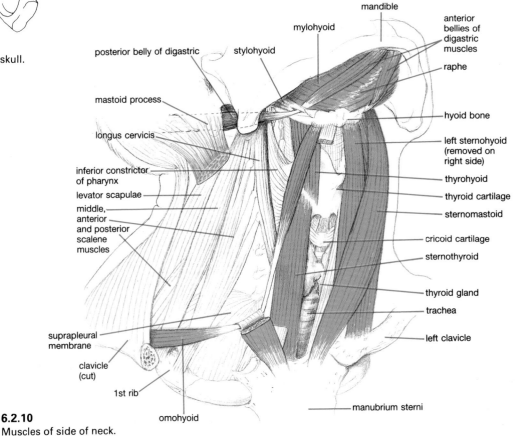

posterior belly of digastric

mastoid process

longus cervicis

inferior constrictor of pharynx

levator scapulae

middle, anterior and posterior scalene muscles

suprapleural membrane

clavicle (cut)

1st rib

omohyoid

stylohyoid

mylohyoid

mandible

anterior bellies of digastric muscles

raphe

hyoid bone

left sternohyoid (removed on right side)

thyrohyoid

thyroid cartilage

sternomastoid

cricoid cartilage

sternothyroid

thyroid gland

trachea

left clavicle

manubrium sterni

6.2.10
Muscles of side of neck.

laterally to insert into the **scalene tubercle** on the medial border of the downward-sloping 1st rib. Now identify **scalenus medius** and **scalenus posterior** which arise from the posterior tubercles of the transverse processes of the cervical vertebrae. Scalenus medius inserts into the upper aspect of the 1st rib posteriorly; scalenus posterior passes down on to the 2nd rib.

Lying superior to the scalene muscles is **levator scapulae** which arises from the posterior aspect of the upper cervical transverse processes and passes downward to insert into the superomedial angle of the scapula, which it elevates.

All the muscles of these groups are supplied segmentally by cervical nerves. Anterior primary rami supply the flexors and scalenus anterior; posterior primary rami supply scalenus medius, scalenus posterior, and all the extensor group.

Identify the **sternomastoid** muscles (**6.2.1**) which lie obliquely at the side of the neck. Each has a tendinous origin from the front of the manubrium and a fleshy origin from the medial part of the clavicle, the long parallel fibres inserting into the mastoid process and superior nuchal line of the occipital bone. It is supplied by the spinal root of the accessory nerve (p. 128). Together the muscles act to protract the head.

Qu. 2C *Sternomastoid muscle on one side may not develop properly, thus remaining short; alternatively, damage to its nerve supply at birth during forceps delivery may paralyse the muscle. Describe the abnormal position (congenital torticollis; twisted neck) that would result from this, or from unilateral contraction of the muscle.*

Identify **trapezius** (**6.2.2**) which takes origin from the external occipital protuberance and the superior nuchal lines on the occipital bone, the nuchal ligament attached to the cervical vertebral spines, and the spines of the thoracic vertebrae. The uppermost muscle fibres insert into the inner margins of the lateral one-third of the clavicle, the acromion process, and the upper aspect of the spine of the scapula; the lower fibres also insert into the upper margin of the spine of the scapula. The upper fibres raise the scapula, whereas the lower fibres on contraction rotate the scapula on the chest wall so that the glenoid fossa (and therefore the shoulder joint) is raised and the arm can be lifted (abducted) above the head (Vol. 1, Chapter 6, Seminar 2). Trapezius is also supplied by the spinal part of the accessory (XI) nerve (p. 128).

Triangles of the neck and their associated fascia (6.2.11)

For ease of description, two superficial 'triangles' of the neck, defined by sternomastoid and trapezius, are recognized. The **anterior triangle** is the area on the anterior aspect of the neck between the two sternomastoid muscles,

bounded above largely by the lower jaw; the **posterior triangle** lies on the side of the neck between the sternomastoid and trapezius and is bounded inferiorly by the middle third of the clavicle. Strong layers of fasica form the roof and floor of these triangles. The **investing layer of (deep) cervical fascia** (**6.2.12**) surrounds the neck between the shoulder girdle and skull. On each side it encloses sternomastoid anteriorly and sweeps laterally around the neck to enclose trapezius posteriorly. Inferiorly, it is attached to the upper border of the manubrium, to the clavicle, acromion, and spine of the scapula; superiorly, to the lower border of the mandible as far as the angle, to the mastoid process, and superior nuchal line. It thereby forms the roof of both the anterior and the posterior triangles. The prevertebral fascia (**6.2.12**), as its name suggests, invests the deep prevertebral muscles of the neck thereby forming the floor of the posterior triangle. It is dense in the midline, where it lies behind the pharynx and oesophagus, but becomes thinner more laterally. Inferiorly, there is a triangular gap in the prevertebral fascia between the lower border of scalenus anterior and the lateral border of the cervical vertebrae (and longus colli) through which the vertebral vessels pass to gain access to the foramina transversaria of the cervical vertebrae (p. 104).

Between the origins of scalenus anterior and the combined scalenus medius and posterior lie the intervertebral foramina. Therefore any cervical nerves, whether they form the cervical or brachial plexuses, must pass into the neck between these groups of muscles and behind the prevertebral fascia. Therefore branches of the cervical plexus which pass to superficial structures in the neck must pierce the prevertebral fascia. Around the combined C5–T1 nerve roots which form the brachial plexus (Vol. 1, Chapter 6, Seminar 8) the prevertebral fascia continues as the **axillary sheath**.

Temporomandibular joints

Take a skull, and disarticulate the mandible. Remind yourself of its major components, in particular the **head of the mandible** which

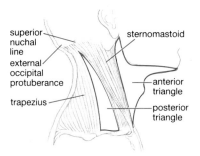

6.2.11
'Triangles' of neck.

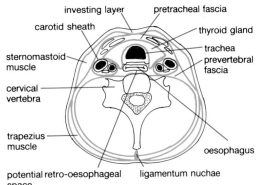

6.2.12
Fascial planes of neck.

forms a rounded bar lying horizontally across the **neck of the mandible** (see **6.1.19**). Articulate the mandible with the skull and identify the **mandibular fossa** in which lie the head of the mandible and its associated disc. The fossa is formed largely from the squamous part of the temporal bone and its zygomatic process anteriorly. The **tympanic plate** of the temporal bone lies posterior to the fossa, separated from the squamous part by the squamotympanic fissure in the depths of which a small piece of the petrous temporal bone often obtrudes. The chorda tympani nerve (p. 123) and a small artery to the middle ear (p. 107) pass through the fissure.

Each temporomandibular joint is a synovial joint, and the left and right joints move together. The head of the mandible is separated from the articular fossa by a fibrocartilaginous articular disc. Examine the **capsule** of the joint: its upper part is thin and elastic; posteriorly, it is attached to the articular fossa just in front of the squamotympanic fissure; anteriorly, it is attached to the apex of the articular eminence on the zygomatic process of the temporal bone; the capsule is attached firmly to the intraarticular disc and, below that, firmly to the neck

of the mandible. The capsule is reinforced laterally by the **temporomandibular ligament** (**6.2.13**). Locate the ligament which passes downward and backward from the zygomatic arch to the lateral side of the neck of the mandible.

Examine the medial aspect of the joint on a prosection. It is reinforced by two extracapsular ligaments: the **sphenomandibular ligament** connects the spine of the sphenoid to the lingula of the mandible; and the **stylomandibular ligament** (a thickening of the investing layer of cervical fascia) connects the styloid process to the angle of the mandible.

Examine the articular disc within the joint: the disc is concavo-convex above to fit the articular eminence and fossa, and concave below to receive the head of the mandible. At its margins, the disc is attached to the capsule, thereby creating two joint cavities, one above and the other below. The posterior part of the disc divides into a fibroelastic upper lamella which attaches to the posterior margin of the mandibular fossa and a fibrous lower lamella which attaches to the back of the condyle. When the mouth opens, a hinge-like movement occurs be-

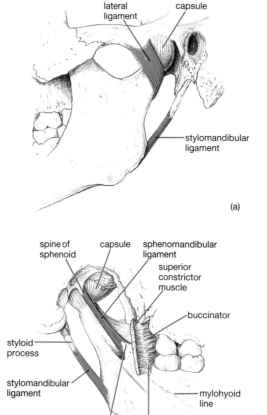

(a)

(b)

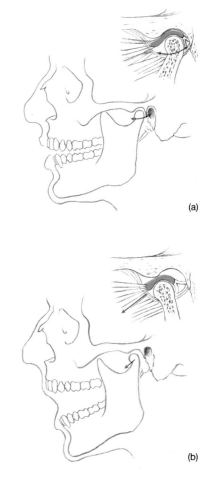

(a)

(b)

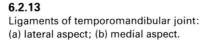

6.2.13
Ligaments of temporomandibular joint:
(a) lateral aspect; (b) medial aspect.

6.2.14
Position of head of mandible with (a) mouth closed and (b) mouth open.

tween the head and the disc (**6.2.14**), and the articular disc and the head of the mandible slide forward together under the articular eminence. Small movements of the jaw, as in talking, are accomplished by a hinge movement between the head of the mandible and disc alone. Because of the firm attachment of the disc to the mandible, the hinge and sliding movements occur in separate compartments of the joint.

Qu. 2D *What movement occurs in the upper cavity of the joint, and what movement in the lower?*

Qu. 2E *When a jaw is dislocated and the head of the mandible is displaced forward beyond the articular eminence, what manoeuvre might be used to reduce the dislocation?*

Qu. 2F *When the jaw is knocked backwards by a blow, which nerve(s) may be damaged? (answer after studying Seminar 12.)*

Muscles of mastication

Examine **masseter** (**6.2.15**) and **temporalis** (**6.2.16**) which you have already felt contracting. **Masseter** arises in several different layers from the zygomatic arch and passes downward and backward to insert into the outer aspect of the angle of the jaw and lower half of the ramus of the mandible. **Temporalis** arises from the side of the temporal fossa and from its covering fascia, and its fibres pass beneath the zygomatic arch to be attached to the medial surface of the coronoid process of the mandible, to its anterior border, and to the anterior border of the ramus of the mandible. Both temporalis and masseter act to close the mouth.

Qu. 2G *Some fibres of temporalis take origin from the temporal bone as far back as the upper aspect of the mastoid process. What action would these fibres have?*

The **lateral pterygoid muscles** (**6.2.17**) are the prime openers of the mouth. They provide the protraction of the head of the mandible necessary for wide opening. Look carefully at their attachments: each arises from the lateral aspect of the lateral pterygoid plate of the sphenoid and from the infratemporal surface of the greater wing of the sphenoid; the fibres pass backward and laterally to be attached to the anterior aspect of the neck of the mandible and the articular disc. **Medial pterygoid** (**6.2.17**) originates on the medial aspect of the lateral pterygoid plate and a tubercle on the adjacent part of the maxilla. The fibres pass downward, backward, and laterally to the medial aspect of the angle of the mandible.

Qu. 2H *What is the action of the medial pterygoid muscles?*

Qu. 2I *What will be the action if the two sets of pterygoid muscles (lateral and medial) act together?*

If the right and left pterygoid muscles contract alternately then a side-to-side ruminant-like movement results.

All these muscles which take part in mastication are innervated by branches of the mandibular (Vth) nerve (p. 121) which emerges through the foramen ovale into the infratemporal fossa. Two other muscles, the anterior belly of the **digastric** muscle and **mylohyoid**, are also attached to the mandible and supplied by the mandibular nerve. Both are considered with the floor of the mouth (p. 57); the anterior belly of digastric has a weak action in opening the mouth.

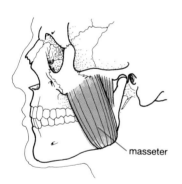

6.2.15
Masseter.

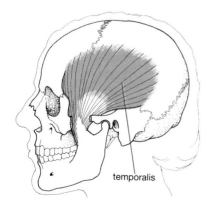

6.2.16
Temporalis.

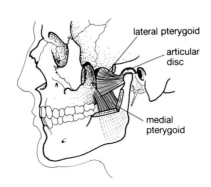

6.2.17
Pterygoid muscles.

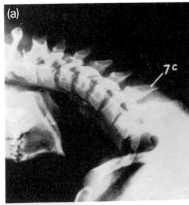

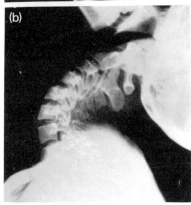

6.2.18

Cervical spine in (a) full flexion and
(b) full extension.

C. Radiology

Examine A–P and lateral radiographs of the
cervical spine (**6.1.29**) and its articulation with
the skull. Identify the component parts of the
atlanto-occipital, atlanto-axial, and intercervical
joints. Note the degree of movement shown by
radiographs taken in full flexion (**6.2.18a**) and
full extension (**6.2.18b**). Examine also the radiograph of the odontoid peg taken obliquely
through the open mouth (see **6.1.30**)

Examine the lateral radiographs of the temporomandibular joint (**6.2.19a,b**) taken with the
mouth closed and widely open. Note how far the
head of the mandible passes forward beneath
the articular tubercle.

Requirements:

Articulated skeleton; separate skull; separate
bones of the cervical spine.

Prosections of the atlanto-occipital, atlanto-
axial, and intercervical joints; prosections
of the neck to show sternomastoid, trapezius, the scalene muscles, the investing
layer of cervical fascia, and the prevertebral
fascia; prosections of the temporomandibular joint, including its articular disc; prosections of the muscles of mastication.

Radiographs of the cervical spine, occipito–
atlanto–axial region and temporomandibular joint.

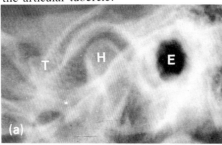 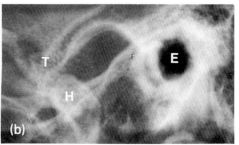

6.2.19

Radiographs of temporomandibular joint: (a) mouth closed; (b) mouth open. Head of mandible
(H); articular tubercle (T); external auditory meatus (E).

Seminar 3

Face, nose, and paranasal sinuses

The **aim** of this seminar is to study, in the living, and in prosections, the face, the nose, and its paranasal sinuses.

Development of the face

(6.3.1, 6.3.2)

The face begins to develop as neural crest cells from the cranial neural folds migrate ventrally and rostrally into the pharyngeal arches and frontonasal region, to form the mandibular, maxillary, and nasal swellings. The mandibular and maxillary swellings surround the primitive mouth (stomodeum) which becomes continuous with the foregut when the buccopharyngeal membrane breaks down. The medial and lateral nasal swellings grow and incompletely surround the nasal placode ectoderm (which is derived from neural ectoderm but comes to lie within the surface ectoderm; p. 20) so that the placode lies at the base of a shallow nasal pit. The two nasal pits deepen until they are each separated from the primitive oral cavity only by a thin **oronasal membrane**. When this membrane breaks down, the nasal cavities become continuous with the oral cavity and primitive pharynx (foregut) but are separated from each other by the nasal septum. During development of the nasal cavities the nasal placode maintains its position close to the forebrain and differentiates to form the olfactory epithelium by extension of neurites into the developing olfactory region of the brain.

The mandibular swellings grow medially and unite in the midline caudal to the stomodeum, forming the primordium of the lower jaw. The medial nasal swellings fuse together in the midline to form the primordia of the median part of the nose, upper lip (philtrum), and upper jaw (primary palate). The maxillary swellings grow medially cranial to the stomodeum and caudal to the developing eye and fuse with the lateral nasal swellings to form the lateral part of the upper lip and jaw and the side walls of the nose. At this line of fusion, which runs upward between the side of the nose and the medial angle of the developing eye, a solid core of cells is formed which later canalizes to form the **nasolacrimal duct**. The mandibular and maxillary swellings also partially fuse to form the cheeks.

Within and on each side of the developing oral cavity shelves of maxillary tissue grow downward, separated by the developing tongue. As the rapidly growing head elongates,

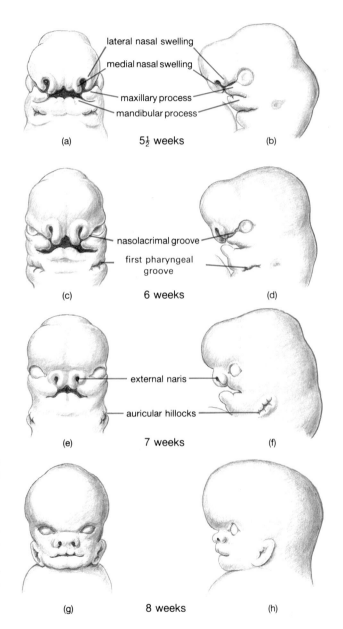

6.3.1
Development of face; frontal and lateral views at (a,b) 5.5 weeks, (c,d) 6 weeks (e,f) 7 weeks, and (g,h) 8 weeks of fetal life.

the tongue descends and the **secondary palatal shelves** meet in the midline and fuse with the primary palate and with each other to form the definitive palate. They also fuse with the ventral edge of the midline **septum**. As a result of

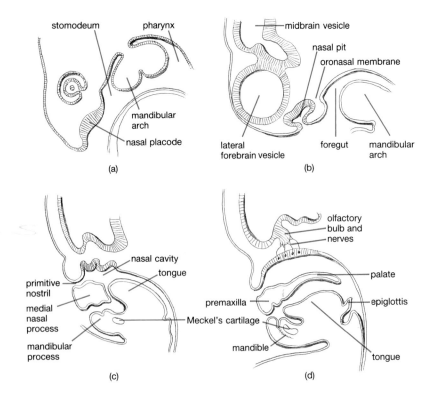

6.3.2.
Development of nose viewed in parasagittal section. (a) Formation of nasal placode; (b) formation of nasal pit and oronasal membrane; (c) breakdown of oronasal membrane; (d) definitive form.

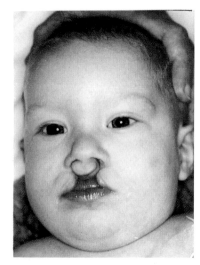

6.3.3
Bilateral lateral cleft of lip.

these fusions, the primitive nasal cavities become divided in the midline and separated from the oral cavity.

If any of these fusions fail owing, among other causes, to genetic or nutritional factors, congenital abnormalities of the face, nose, lips, and palate occur. Lateral clefts of the lip (**6.3.3**) (due to failure of maxillary and medial nasal swellings to fuse) and clefts of the palate are the most common; they cause problems with suckling and later inefficient eating and speaking. To restore adequate function, the abnormalities should be corrected by surgery as soon as possible.

Qu. 3A *If the mandible failed to grow properly so that the child was born with a small lower jaw, what other anomaly might be causally associated?*

The development of the tongue is described on p. 55

A. Living anatomy

Examine your partner's face. Look at the texture and colour of the skin; note any moisture due to sweat or oily secretions from the sebaceous glands. You may also see freckles, pigmented or non-pigmented moles, or infected sebaceous glands or scars caused by acne. Note any sexually dimorphic differences in the hair line of the scalp and the distribution of hair on the face. Examine the eyebrows and note whether the density of hair between the medial and lateral aspects of the eyebrows is similar. Note also any coarse hairs (vibrissae) in the

nostrils and external auditory meatus which protect these orifices.

Are there any blood vessels to be seen on your partner's face or neck? If so, are they pulsating? Palpate the area immediately in front of the external auditory meatus: over the root of the zygomatic arch it is often possible to feel the pulsation of the superficial temporal artery. Get your partner to clench his/her jaws tightly and feel the contraction of masseter at the angle of the jaw: palpate its anterior margin at the inferior edge of the body of the mandible where the facial artery should be easily felt as it crosses bone.

Examine the lips: they appear pink because the stratified squamous epithelium that covers them is non-keratinized and the vascular dermis can be seen.

Qu. 3B *Are your partner's cheeks pink or pale? If the cheeks are pigmented can you tell whether or not they are pink?*

The face is extremely mobile and a vital signalling device. Can you tell from the facial appearance whether your partner is relaxed or tense? Look at the faces of other people in the room. Get your partner to 'look' happy, angry, sad, or show any other emotion. As you study the facial muscles, consider the role they play in different expressions (muscles moving the eyes will be studied in Seminar 8; p. 90). Ask your partner to close his/her eyes gently, and then to 'screw them up'.

Qu. 3C *What difference do you note?*

Using a skull and your partner, examine again the bones of the face (see p. 31).

Qu. 3D *At which sites on the skull are the facial bones buttressed to the cranium?*

The **nose** has two major functions: it is an olfactory sensory organ; and a respiratory passage that warms, filters, and moistens the incoming air. Olfactory receptor functions are considered on p. 115.

The **nose** in coronal section is triangular in shape, and divided by a midline septum into two passages which extend from the nostrils (**external nares**) to the posterior openings into the nasopharynx (**internal nares**). Insert a nasal speculum into one and then the other nostril, gently opening them to see the **conchae** which curve inward and downward from each side wall. The three conchae more than double the internal surface area of the nose.

Palpate the nasal bones, then the cartilaginous parts of the nose, noting the mobility of the latter.

B. Prosections of the face and nose

Muscles of facial expression (6.3.4)

The facial muscles, which all receive motor innervation from the facial (VIIth) nerve (p. 122), surround and control the apertures of the mouth and nose and eyes and, in some species, control the position of the external ears. They support the angles of the eyes and mouth, thereby preventing secretions from running on to the face, while buccinator prevents food from collecting between the teeth and gums during chewing. By altering the facial expression they add subtlety of meaning to the spoken and unspoken word.

Identify on a prosection the muscles labelled in **6.3.4**, all the time making reference to the living faces around you.

Orbicularis oculi consists of two main parts. The flat **orbital** part consists of concentric fibres which encircle the orbit. Although most are attached to the overlying skin, some are inserted into a small area of bone on the medial side of the orbit — these 'screw up' the eyes. The fine **palpebral** fibres pass across the eyelids from the medial palpebral ligament (and adjacent part of the medial wall of the orbit) to interdigitate and form the lateral palpebral raphe (attached to the lateral wall of the orbit) — these close the lids as in sleep or blinking. In addition, a few fibres attached to the lacrimal sac form a **lacrimal** part which may help to dilate the sac and also keep the lacrimal puncta in contact with the eyeball.

Now examine **orbicularis oris** which surrounds the mouth in a sphincter-like manner. Many of the other facial muscles, particularly **buccinator**, and muscles that raise (levator) or depress (depressor) the angles of the mouth, blend with these circular fibres. At the lateral angle of the mouth some fibres run directly into the upper or lower lip while others cross, to form a chiasma.

Buccinator (**6.3.5**), the muscle of the cheek, is a facial muscle which contracts during chewing, thereby preventing food from collecting between the cheek and the gums; it also raises the pressure of air expelled through the mouth as in whistling or playing wind instruments. It arises posteriorly from the fibrous **pterygomandibular raphe**, which is attached between the pterygoid hamulus and the mandible just behind the last molar tooth (retromolar fossa), and from the alveolar margins of the mandible and maxilla adjacent to the molar teeth. Its fibres pass forward to merge with those of orbicularis oris.

Qu. 3E *What disabilities would occur if the nerve supply to the facial muscles on one side were damaged?*

The **scalp** covers the top of the skull and, except in many older men, is covered with hair. The dermal layer, through which the neurovascular supplies run, contains much fibrous tissue which in scalp lacerations prevents constriction of the vessels and leads to copious bleeding. The dermis is firmly attached to an underlying aponeurosis (expanded tendon) which connects two pairs of muscles: **frontalis**, which attaches into the skin of the forehead and the supraorbital ridge, and **occipitalis** which attaches to the highest nuchal line on the occiput. Deep to the aponeurosis is a layer of loose areolar connec-

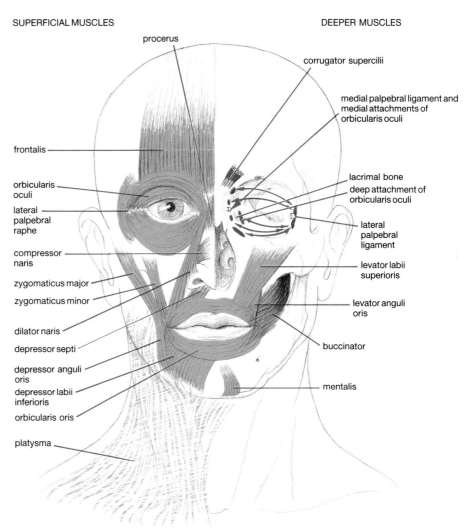

6.3.4
Muscles of facial expression.

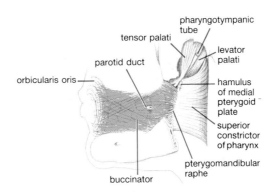

6.3.5
Buccinator.

tive tissue which serves as a bursa and which permits the aponeurosis to move freely on the periosteum of the cranial vault. Deep infections of the scalp lead to marked swelling in this layer.

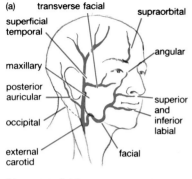

(a)
transverse facial
supraorbital
superficial temporal
maxillary
angular
posterior auricular
occipital
superior and inferior labial
external carotid
facial

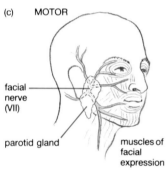

(b)
superficial temporal
supraorbital
communicating
posterior auricular
occipital
retro-mandibular
facial
external jugular
labial
internal jugular
subclavian
anterior jugular

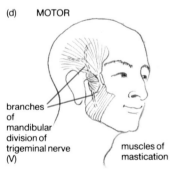

(c) MOTOR
facial nerve (VII)
parotid gland
muscles of facial expression

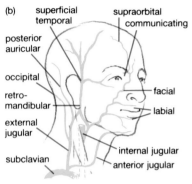

(d) MOTOR
branches of mandibular division of trigeminal nerve (V)
muscles of mastication

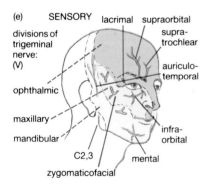

(e) SENSORY
lacrimal
supraorbital
divisions of trigeminal nerve: (V)
supra-trochlear
auriculo-temporal
ophthalmic
maxillary
mandibular
infra-orbital
C2,3
mental
zygomaticofacial

6.3.6
(a) Arteries, (b) veins, and (c–e) nerves of the face.

Blood supply of the face and scalp

Examine a prosection of the face and neck (**6.3.6**). Locate the **facial artery** which arises from the **external carotid artery** behind the angle of the mandible. In the first part of its course the artery loops upward behind the mandible and submandibular gland, then crosses the inferior border of the mandible to gain the face. Trace the artery upward through the facial muscles towards the medial angle of the eye and locate branches to the upper and lower lips, nose, and cheeks. Locate the two terminal branches of the external carotid artery: the **maxillary artery** which passes deep to the mandible and supplies deep aspects of the face; and the **superficial temporal artery** which passes upward over the zygomatic arch to supply the side of the scalp. Locate the **transverse facial artery**, a branch of the superficial temporal that runs across the face parallel to the parotid duct. Small arteries (supraorbital and supratrochlear), derived from the ophthalmic branch of the internal carotid artery, emerge from beneath the upper margin of the orbit to supply the anterior part of the scalp. Locate also the **posterior auricular** and **occipital** branches of the external carotid artery that supply the posterior part of the ear and scalp. The face and scalp have a plentiful arterial supply with many anastomoses (see **6.10.6**) which occur between the vessels of the two sides and between deep and superficial vessels.

Examine a prosection of the **veins** of the face and identify the branches labelled in **6.3.6**.

Lymphatics of the face and scalp run with the superficial veins and drain to the ring of submental, submandibular, parotid, mastoid, and occipital nodes at the upper extremity of the investing layer of cervical fascia. From there they drain mostly to deep cervical nodes (p. 110).

Innervation of the face and scalp
(**6.3.6**)

Sensation to the face is supplied, almost exclusively, by the **trigeminal (Vth cranial) nerve** (p.117). Note now that the trigeminal nerve has three major divisions: the **ophthalmic division** which supplies the 'visor' area of the face, midline of the nose, upper eyelid, and scalp; the **maxillary division** which supplies the lateral aspect of the nose, the cheek, lower eyelid, and part of the temple; and the **mandibular division** which supplies the lower lip, chin, and a strip of skin which runs upward from the angle of the jaw to the vertex of the skull, including the tragus and upper part of the pinna. The skin over the angle of the jaw is supplied from the **anterior primary ramus of C2**, and neck skin is supplied by branches of the anterior primary rami of **C2**, **C3**, and **C4** (p. 129). Behind the vertex, the scalp and neck are supplied by the **posterior primary rami of C2 and C3** (scalp) and **C3–C7** (neck).

Qu. 3F *If the ophthalmic division of the*

trigeminal nerve were damaged, what might happen to the eye?

The **motor supply** to the **muscles of facial expression** (including buccinator) is derived from numerous branches of the **facial (VIIth cranial) nerve** (p. 122) which emerge from the parotid gland on to the face (**6.3.6**). Its cervical branch, which supplies platysma, hangs below the angle of the mandible.

The nose

Examine the nasal cavity of a head which has been sectioned in the sagittal plane (**6.3.7**); it extends from the nostrils (external nares) to the posterior nares (internal nares) which open on to the nasopharynx. The roof of the nose is narrow and slopes downward both anteriorly and posteriorly. With a skull, remind yourself of the skeleton of the nose. Its **floor** is the upper surface of the hard palate, formed from the maxilla and palatine bones; its **lateral walls** are formed by the medial walls of the maxillary and ethmoid air sinuses and, more posteriorly, by the medial pterygoid plate of the sphenoid; its **roof** is formed from the nasal bones anteriorly, cribriform plate of the ethmoid, and the undersurface of the body of the sphenoid; its two halves are separated by the **nasal septum** formed from a septal cartilage anteriorly, and vertical plates of the ethmoid and vomer posteriorly.

Qu. 3G *How might you detect a fracture of the anterior part of the base of the skull by inspection of the nose?*

Identify the three **conchae** which project from the side wall of the nose; each overhangs a defined space (meatus). The **inferior concha** (turbinate) is a separate bone which articulates with the medial wall of the maxillary antrum; it overhangs the **inferior meatus** through which the respired air largely passes. The **middle concha** that overhangs the **middle meatus** is shorter than the inferior concha; it is part of the ethmoid bone. The **superior concha** is small and lies superoposteriorly above the **superior meatus**; it too is a part of the ethmoid bone.

A highly vascular **ciliated epithelium** lines all the nasal cavity except the roof and superior conchae (which is covered with olfactory epithelium). The epithelium contains many goblet cells which, with the serous secretions of submucous glands, cover the interior of the nose with a blanket of moist mucus. Just beneath the surface is a cavernous plexus of blood sinuses with arteriovenous anastomoses which open or close to control blood flow. This arrangement, coupled with turbulence of the air produced by the arrangement of the conchae, enables environmental air to be adjusted to body temperature and humidity by the time it reaches the nasopharynx. In response to increased stimulation of the autonomic nervous system, or proteins such as pollen to which an individual is sensitized ('hay-fever'), the mucosa becomes swollen and mucus secretion is increased.

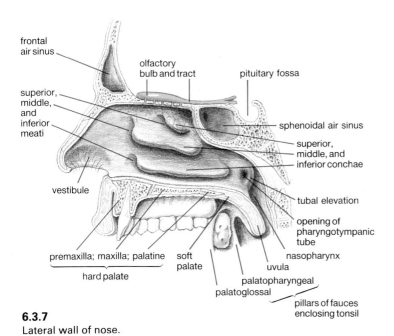

6.3.7
Lateral wall of nose.

Note coarse hairs, '**vibrissae**', in the **vestibule** of the nose just inside the external nares; these filter coarse particles from the incoming air. Particles which escape are trapped by the mucus. Cilia, which cover the surface of the epithelial cells, sweep the mucus at 5–10 mm/min to the nasopharynx, to be swallowed. Most bacteria are destroyed either by immunoglobulins in the nasal mucus or later by the acid secretions of the stomach.

Drainage of paranasal sinuses into the nose

Now turn to a specimen in which the conchae have been removed to expose the meatuses (**6.3.8**). Explore the inferior meatus with a blunt probe. Find the **nasolacrimal duct**, which carries the lacrimal secretions from the **lacrimal sac** in the medial wall of the orbit (p. 93) to the nose. In the middle meatus locate a bony protuberance known as the **ethmoidal bulla** on to which numerous ducts from the anterior and middle ethmoidal air cells open. The **duct of the frontal air sinus** opens close to its anterior extremity, and the **opening of the maxillary antrum** drains posteriorly. Now search for the **posterior ethmoidal air sinuses** which open into the superior meatus and, above and posterior to the superior concha, the **sphenoidal recess** into which drains the large **sphenoidal air sinus**. Explore the relationship of the sphenoidal air sinus with the pituitary fossa which is formed by its roof.

Examine the **septum** of the nose (**6.3.9**) which is formed from cartilage and bone. Its posterior part is bony, formed by the vomer, the perpendicular plate of the ethmoid superiorly, and a small upward projection of the maxillary bones inferiorly. Anteriorly, the septum consists of a quadrilateral cartilaginous plate which articulates with the septal bones and also with the

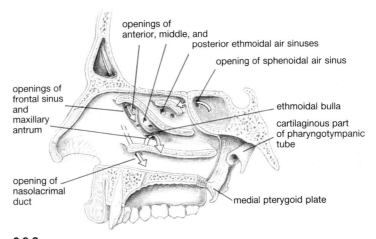

6.3.8
Lateral wall of nose with conchae removed.

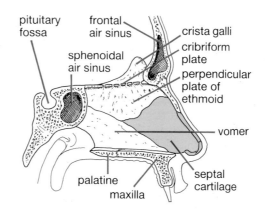

6.3.9
Nasal septum.

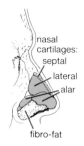

6.3.10
Nasal cartilages.

alar cartilages of the nostrils (6.3.10). The very flexible tip of the nose is formed by septal extensions of these nasal cartilages.

Qu. 3H *Not infrequently, and especially after the nose has been broken (6.3.13), the septum becomes deviated to one side. What dysfunction might this cause?*

Paranasal air sinuses

Examine a coronal section of the skull in which the ethmoidal and maxillary air sinuses can be seen (6.3.11) and look carefully at the ethmoid (6.3.12). The ethmoid appears complex until it is realized that it is formed of two sets of honeycombed air cells which form the medial wall of the orbit laterally and the lateral wall of the nose, medially. The two sets of ethmoid cells are bridged dorsally by a shelf of bone which is perforated on either side of the midline (the cribriform plate) to allow the neural fibres of the olfactory epithelium to pass into the cranium. Intersecting the midline is a perpendicular plate of bone which extends upward into the cranial cavity as the crista galli, and downward into the nasal cavity to form part of the septum of the nose. The ethmoidal air cells

open into the middle and superior meatuses of the nose. Since the ethmoid air cells form part of the medial wall of the orbit it is therefore very thin and fragile and liable to injury (6.3.13). Surgical access to the pituitary gland can be gained either through the ethmoid bones on the medial wall of the orbit or through the roof of the nasopharynx beyond which is the sphenoidal sinus and then the pituitary fossa. Examine the frontal air sinuses which lie in the frontal bone on either side of the midline and drain into the middle meatus. They develop at about four years of age and may be absent or poorly developed on one or both sides. Examine the maxillary air sinuses, which are situated in the body of the maxillary bones on either side of the face and are roughly pyramidal in shape with the base of the pyramid forming the lower part of the lateral aspect of the nose (6.3.11, 6.3.16). Pass a probe into the sinus from the ostium leading to the middle meatus of the nose. The other three walls of the pyramid are orientated in an orbital, facial, and infratemporal direction. The floor of the maxillary sinus is formed by the alveolar process of the maxilla and roots of the second upper premolar (covered by mucous membrane) may be seen projecting into the sinus on one or other side.

The paranasal sinuses are all lined by ciliated cuboidal epithelium which helps to keep the spaces properly drained. Their presence enlarges but lightens the facial skeleton and gives resonance to the voice.

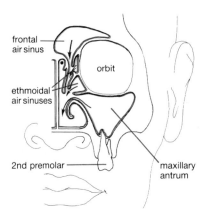

6.3.11
Frontal, ethmoidal, and maxillary sinuses and their drainage into the nose.

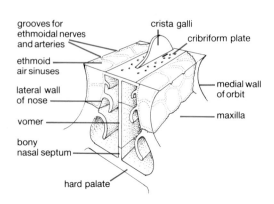

6.3.12
Ethmoid bone (diagrammatic).

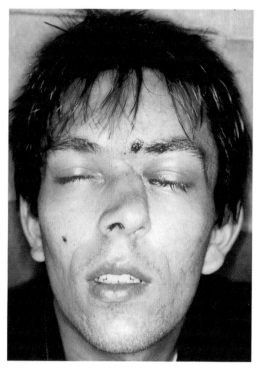

6.3.13
Fracture of ethmoid bone. Note displacement at root of nose.

Qu. 3I *If an abscess develops in the root of an upper premolar tooth, to where might the infection spread?*

Qu. 3J *How might a maxillary sinus be drained surgically if it were full of stagnant, possibly septic, mucus?*

Blood supply of the nose (6.3.14)

The **arterial supply** of the nose is derived mainly from the **maxillary** artery (from the external carotid artery), through branches which accompany the nerve supply from the branches of the maxillary (Vth) nerve (see p. 129). **Ethmoidal** branches of the ophthalmic artery (from the internal carotid artery) to the ethmoid air cells supply the upper anterior part of the nose, and branches of the **facial** artery supply the vestibule. An anastomosis of these arteries is formed on the cartilaginous part of the septum (Little's area) and damage to this area can result in serious bleeding.

Qu. 3K *If a blow to the nose results in severe bleeding, what simple measure could stop the bleeding?*

Venous drainage from the rich submucous plexus passes with the branches of the maxillary artery to the pterygoid plexus which drains into retromandibular veins and communicates with the cavernous sinus; and with ethmoid vessels to the cavernous sinus via ophthalmic veins. Others join the facial vein or pass into the cranial cavity through the cribriform plate.

Qu. 3L *Why is the venous drainage of the nose so rich?*

Lymphatic vessels from the anterior part of the nasal cavity run with the anterior facial vein to enter the submandibular group of glands; from the remainder, they run to lymphatic tissue around the nasal and oral parts of the pharynx and thence to upper deep cervical nodes (p. 110).

Innervation of the nose (6.3.15)

The interior of the nose is supplied by nerves which convey to it sensory, parasympathetic (secretomotor), and sympathetic fibres. Most of the sensory fibres are supplied by the **maxillary division of the trigeminal (Vth) nerve** through branches which pass through but do not synapse in the **sphenopalatine ganglion**. Locate the sphenopalatine ganglion which hangs beneath the maxillary nerve in the sphenopalatine foramen covered by mucous membrane of the posterior aspect of the middle meatus. It gains **parasympathetic** (preganglionic) fibres indirectly from the facial (VII) nerves which synapse in the ganglion; and **sympathetic** (postganglionic) fibres from the internal carotid plexus which do not synapse. Locate the **palatine nerves**, branches from the ganglion which pass downward through a bony canal in the palatine bone to emerge on the oral surface of the hard palate. Trace the **long sphenopalatine nerve** which passes from the ganglion, across the roof of the nose, and then diagonally downward and forward along the nasal septum to emerge through the **incisive foramen** on to the roof of the mouth. With other small branches from the sphenopalatine ganglion to the upper posterior part of the lateral walls and septum the maxillary nerve supplies common sensation to most of the side walls and septum of the nose. The anterior aspect of the nasal cavity, however, is supplied by the **internal nasal** branch of the ophthalmic division of the trigeminal nerve via its anterior ethmoid branch. The anterior ethmoid terminates by passing between the nasal bone and cartilage to form the **external nasal** nerve.

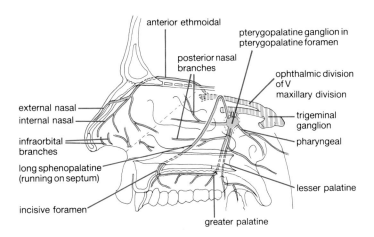

6.3.15
Nerve supply of nose and palate. Demarcation between maxillary and ophthalmic divisions of trigeminal nerve (dashed line).

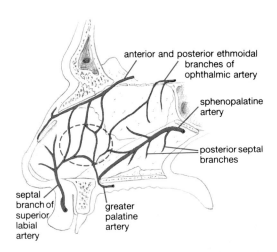

6.3.14
Arterial supply of nose. Area of anastomosis (ringed).

C. Radiology

Examine a frontal radiograph (6.3.16) and define the bones which form the facial skeleton. Identify the maxillary antra (M) on either side of the nasal cavity, the frontal sinuses (F) above and medial to the orbits, and the ethmoid air cells (E) which in frontal radiographs are partly superimposed on the nasal cavity. Note the septum of the nose, which may be deviated from the midline.

Examine the horizontal CT (6.3.17) through the maxillary antra. Note the inferior conchae on the lateral walls of the nasal cavity. Immediately posterolateral to the antra are the coronoid processes of the mandible and, behind this, the medial and lateral pterygoid muscles of both sides. Identify the pterygoid plates of the sphenoid on either side of the posterior nasal aperture.

Examine the horizontal CT at the level of the ethmoid air cells and sphenoid sinus (6.3.18); at this higher level the nasal cavity is very narrow, being limited to a slit on either side of the midline septum. The ethmoid air cells are complex and separate the nasal cavity from the orbit; immediately posteriorly is the large air space of the sphenoid sinus (in this individual divided by a midline bony septum). Locate the foramen lacerum (into which the internal carotid artery emerges from the carotid canals) on either side of the posterior limit of the sphenoid sinus.

The oblique CT (6.3.19) shows the posterior part of the maxillary antra and the sphenoid sinus. Note that the maxillary antra are larger posteriorly than anteriorly (compare with 6.3.17) and in this section all three nasal conchae can be seen clearly.

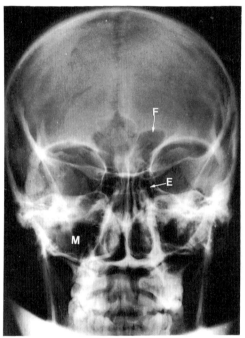

6.3.16
Paranasal sinuses: frontal (F); ethmoid (E); maxillary (M).

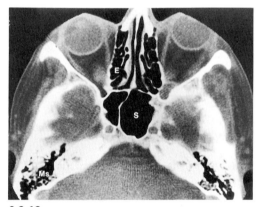

6.3.18
Horizontal CT through ethmoid (E), sphenoid (S), and mastoid (Ms) air sinuses.

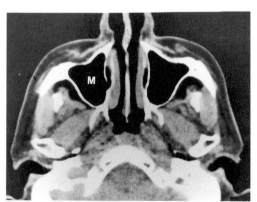

6.3.17
Horizontal CT through maxillary antra (M).

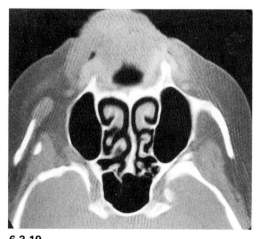

6.3.19
Oblique CT through nose, maxillary, and sphenoidal air sinuses.

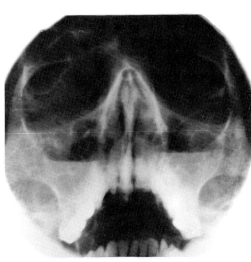

6.3.20
See **Qu. 3M**.

Qu. 3M *Examine radiograph* **6.3.20**. *What abnormality can you detect?*

Requirements:

Skull and skull cap; midline sagittal section of a skull with nasal septum intact; sectioned skull showing opening of maxillary sinus into the nose.

Prosections of muscles of facial expression; blood supply to the face; nerve supply (motor and sensory) to the face; prosections of the nose showing intact conchae; conchae removed to show meatuses; blood and nerve supply.

P–A, lateral, and basal radiographs of the skull, radiographs of paranasal sinuses; CTs.

Nasal speculum and disinfectant.

Seminar 4

Mouth

The **aim** of this seminar is to study the mouth, including the tongue, the floor of the mouth, the hard and the soft palate, and the salivary glands, in the living and by the use of prosections.

Development of the mouth, tongue, and palate
(**6.4.1, 6.4.2**)

The oral cavity is derived partly from the ectodermal **stomodeum** and partly from the most cranial part of the endoderm-lined foregut. These two components are initially separated by the **oropharyngeal membrane** which breaks down and disappears without trace. The epithelium of lips, gums, and tooth enamel is formed from ectoderm; the epithelium of the tongue is derived from endoderm. With the formation of the 1st pharyngeal arch, the stomodeum is bounded by the frontonasal, maxillary, and mandibular processes. Surface ectoderm covers and invades the free margins of these processes to form a labiogingival groove which separates the lips and cheeks from the gums and jaws, and a dental lamina within them.

Within the floor of the primitive mouth a midline tongue-bud swelling — the **tuberculum impar** — develops, with two lateral **lingual swellings** on the inner aspect of the mandibular arch. The lingual swellings meet anteriorly, and grow over and fuse with the tuberculum impar to form the anterior two-thirds of the tongue around which a sulcus deepens to separate the tongue from the gums. The posterior one-third of the tongue forms from the midline **hypobranchial eminence**, derived largely from the 3rd pharyngeal arch. As these components grow, 2nd pharyngeal arch tissue is excluded from the tongue. The junction is marked by the V-shaped **sulcus terminalis**. Common sensation to the anterior two-thirds of the tongue is derived from the mandibular (V; 1st arch) nerve; taste fibres are derived from a pre-trematic branch (chorda tympani; p. 123) of the facial (VII; 2nd arch) nerve. Common sensation and taste to the posterior one-third of the tongue are both supplied by the glossopharyngeal (IX; 3rd arch) nerve. Caudal to the tongue, the vallecula and epiglottis are derived from the 4th arch and innervated by the superior laryngeal branch of the vagus (X; 4th arch) nerve.

Muscles of the tongue are formed from occipital myotomes which migrate into the floor of the mouth, bringing with them innervation from the hypoglossal (XII) nerve.

At the apex of the sulcus terminalis a small pit, the **foramen caecum**, forms in the midline. From this pit the rudiment of the thyroid gland grows down through the tongue toward the larynx and trachea (p. 76).

The roof of the mouth is formed by the palate (**6.4.2**). Palatal shelves of the maxillary processes (1st arch) fuse above the tongue to form

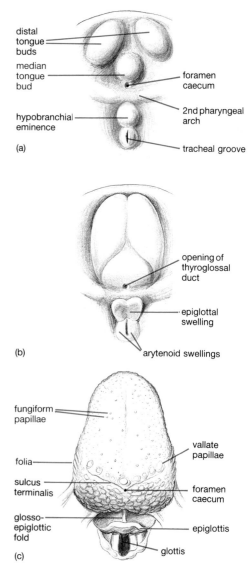

6.4.1

Development of tongue; progressive stages (a–c).

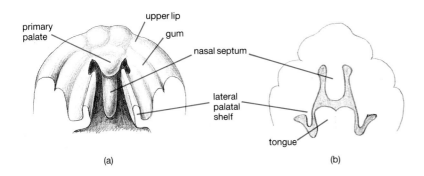

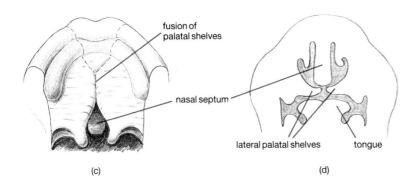

6.4.2
Development of palate. (a,c) Roof of mouth in 7- and 9-week embryos; (b,d)
coronal sections to show changing relationships of palatal shelves and tongue.

the hard palate which is supplied by the maxillary (V; 1st arch) nerve; more caudally, contributions from the 3rd arch form the soft palate and are supplied by the glossopharyngeal (IX; 3rd arch) nerve. Failure of the shelves to fuse results in clefts of the palate, which may be associated with clefts of the lip (**6.4.3**).

A. Living anatomy

Examine your own mouth and that of your partner (**6.4.4**) by use of mirror and torch where necessary. The cavity of the mouth extends backward from the lips to an arch, the oropharyngeal isthmus, which is made up of the soft palate, the upper surface of the tongue, and the palatoglossal fold which extends between the palate and the tongue (**6.4.4**).

Note the red, shiny appearance of most of the oral mucosa and contrast this with the pale pink, stippled appearance of the gums (gingivae)

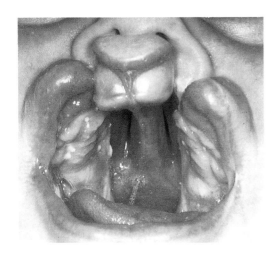

6.4.3
Cleft palate associated with bilateral, laterally cleft lip; presence of nasal septum, absence of palate.

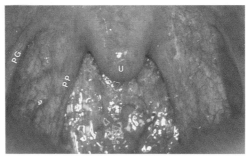

6.4.4
Open mouth showing oropharyngeal isthmus. Palatoglossal fold (PG); palatopharyngeal fold (PP); uvula (U).

which should be adherent to the base of the crowns of the teeth.

Each tooth has an **occlusal** (biting) surface made up of **cusps** (tubercles) separated by fissures; identify the labial and buccal surfaces which lie adjacent to the lips and tongue; and mesial and distal surfaces which face the tooth 'in front' and 'behind'. Count the number of incisor, canine, premolar, and molar teeth in both jaws.

Ask your partner to close his mouth whilst baring his teeth and note whether the incisors of the upper jaw lie in front (usual) or behind the lower incisors. Anomalies of positioning of the lower jaw in relation to the upper jaw result in problems with chewing and the appearance of the face.

Examine the surface of your partner's protruded tongue.

Qu. 4A *Why does the surface appear to be rough, compared with the inner surface of the cheek?*

Identify the V-shaped sulcus terminalis on the upper surface of the tongue and, just anterior to it, the large **vallate papillae**. Note that very little of the posterior one-third of the tongue can be seen — indeed, when chewing food the soft palate is held in contact with the tongue at the sulcus terminalis so that the anterior two-thirds are in the mouth, and the posterior one-third forms the anterior wall of the oropharynx. Also, when chewing, food is kept between the teeth by the combined actions of the tongue and the cheek muscle, buccinator. Demonstrate the enormous range of movements of which the tongue is capable; and demonstrate buccinator by pursing your lips and blowing out, as in playing a trumpet.

Qu. 4B *What type of movements can the tongue execute?*

Raise the tip of the tongue and examine its under-surface. Identify the midline **frenulum**, running from the floor of the mouth to the under-surface of the tongue. Lingual veins draining the tongue are clearly seen on its under-surface.

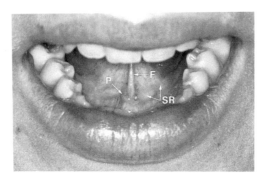

6.4.5
Open mouth showing under-surface of tongue.
Frenulum (f); sublingual ridge (SR);
submandibular papilla (P).

Examine the floor of the mouth (**6.4.5**); this is covered with mucous membrane which extends from the side of the tongue over the floor of the mouth to the inner aspect of the lower jaw and gums. Place your left thumb inside your mouth between the tongue and the lingual surface of the mandible on the right side; now press your index finger upwards beneath the mandible and determine the thickness of the floor of the mouth.

On the floor of the mouth, on either side of the frenulum, locate the **sublingual papillae** on to which open the ducts of the submandibular salivary glands. If these are difficult to find, squeeze a little lemon juice on to the tip of the tongue and look again.

Qu. 4C *Why should lemon juice affect the secretion of saliva?*

Lateral to the openings of the submandibular ducts identify the **sublingual ridges** which indicate the position of the underlying **sublingual salivary glands**. The numerous small ducts of these glands open on to the surface of the ridges.

Qu. 4D *Where does your dentist place the sucker when filling a tooth in your lower jaw?*

Examine the hard palate and its posterior continuation, the soft palate, with its midline **uvula**. Feel how firmly the mucosa is attached to the bone of the hard palate.

Qu. 4E *Why is the mucosa of the hard palate so firmly attached to bone?*

Beyond the uvula identify the posterior wall of the oropharynx This is more easily seen if the soft palate is elevated, traditionally by asking the person being examined to open the mouth wide and say 'aah' (contraction of levator palati raises the soft palate).

Examine the aperture between the mouth and oropharynx — the **oropharyngeal isthmus** (isthmus of fauces). Passing downward from the soft palate on either side of the isthmus are the anterior and posterior **pillars of the fauces**. These are formed by the **palatoglossal** (anterior) and **palatopharyngeal** (posterior) **arches**, each of which comprises a muscle covered by mucous membrane and running from the soft palate to either the tongue or pharynx (palatoglossus or palatopharyngeus). During chewing, the muscles pull the soft palate downward on to the back of the tongue and thereby close the oropharyngeal isthmus. At the onset of swallowing (p. 68) they relax.

Gently touch either the posterior one-third of the tongue or the under-surface of the soft palate with a wooden tongue depressor.

Qu. 4F *What happens, and why? (see p. 143)*

Between the palatopharyngeal and palatoglossal folds lies a variable mass of lymphoid tissue, the **palatine tonsil**. This is larger in children and is frequently infected and inflamed during upper respiratory tract infections. It is part of an incomplete ring of diffuse lymphoid tissue which forms the first defence against bacteria entering through the mouth. The ring extends around the oropharynx (**pharyngeal tonsil** or **adenoid**), continues on either side with the rather discrete palatine tonsil, and extends on to the base of the tongue ('lingual tonsil').

Place the tip of your tongue on the inside of your cheek opposite the 2nd upper molar tooth and locate the small elevation which marks the opening of the duct of the parotid salivary gland. Try to locate the duct opening in your partner's mouth.

B. Prosections of floor and roof of the mouth, tongue, and salivary glands

Floor of the mouth

Review the mandible and hyoid bone (p. 32) and then examine prosections in which the head has been hemisected sagitally and the floor of the mouth dissected from the lateral aspect (**6.4.6a**). Locate **mylohyoid**, a thin muscular gutter which forms the mobile floor of the mouth. Its fibres pass medially from the mylohyoid lines on the inner aspect of the mandible. Anterior fibres form a midline raphe between the mandible and the hyoid; posterior fibres insert into the body of the hyoid bone leaving a free posterior border (**6.4.7**). Superficial to mylohyoid lies the **digastric muscle**; its **anterior belly** passes from the digastric fossa on the inner surface of the mental symphysis downward and backward to an intermediate tendon which is attached to the root of the greater horn of the hyoid bone by a fibrous loop; its **posterior belly** extends backward and upward to a (digastric) notch on the deep surface of the mastoid process. Both mylohyoid and digastric raise the floor of the mouth and hyoid bone during swallowing (p. 68); digastric can depress the mandible in forced opening of the mouth. Identify **stylohyoid** which runs from the styloid process to the junction of the body and greater horn of the hyoid, its tendon dividing around the intermediate tendon of digastric. Find the small nerve and artery to mylohyoid and the anterior belly of digastric which leave the inferior alveolar nerve (mandibular V) and maxillary artery respectively at the posterior border of the mylohyoid, and run forward superficial to mylohyoid close to its attachment to the mandible. The posterior belly of digastric and stylohyoid are supplied by the facial nerve (VII).

Tongue

Examine first a midline section of the tongue. It consists of bundles of **intrinsic** muscle fibres which run longitudinally, transversely, and vertically (**6.4.8**); these blend with **extrinsic** muscles which both move the tongue and anchor it to surrounding bones (**6.4.6**). At the anterior

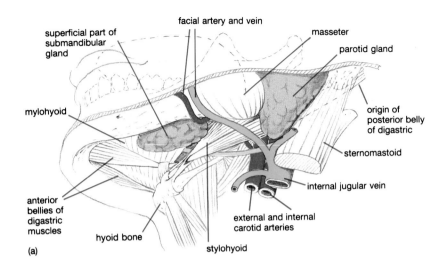

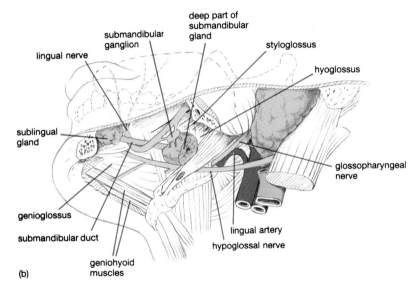

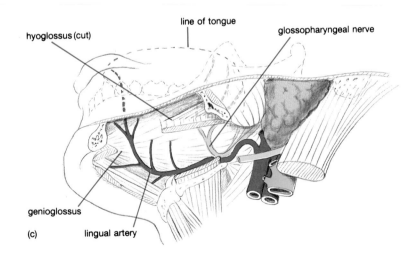

6.4.6
Progressively deeper dissections (a), (b), and (c) of side of mouth.

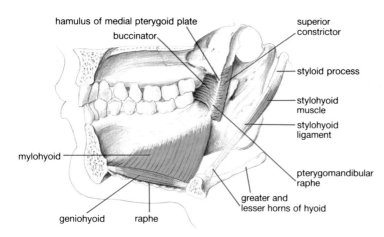

6.4.7
Muscles attached to hyoid bone.

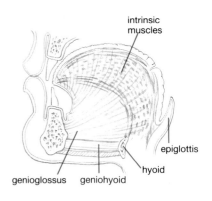

6.4.8
Midline section through tongue.

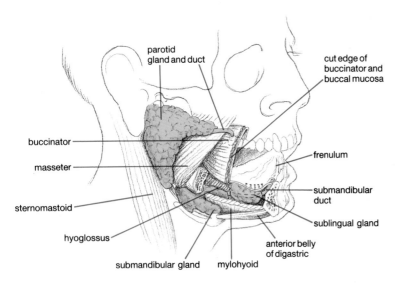

6.4.9
Salivary glands

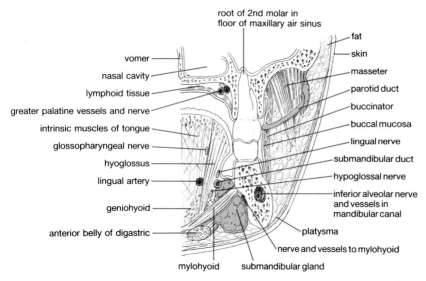

6.4.10
Coronal section through mouth.

aspect of the root of the tongue locate the genial tubercles on the inner surface of the mental symphysis. From the superior genial tubercles arise fibres of **genioglossus** which fan out centrally throughout the length of the tongue from its tip to its root, blending with the intrinsic fibres; it draws the tongue forward and protrudes its tip.

On the side of the tongue locate three extrinsic muscles which interdigitate there (**6.4.6b**). **Hyoglossus** arises from the body and greater horn of the hyoid bone and its parallel fibres pass upward into the side of the tongue to merge with styloglossus and palatoglossus. **Styloglossus** arises from the styloid process and runs downward and forward to the side of the tongue. **Palatoglossus** arises in the soft palate and passes downward in the palatoglossal fold to the side of the tongue; it is essentially a muscle of the palate.

Qu. 4G *What are the actions of hyoglossus and styloglossus?*

From the inferior genial tubercles the 'strap' muscle **geniohyoid** passes backward and downward above mylohyoid to insert into the body of the hyoid.

Qu. 4H *Bearing in mind the attachments of the tongue and its support by mylohyoid, what would be the effect of a bilateral fracture of the body of the mandible which caused the anterior part of the bone to become separate and loose (as can result from heavy blows to the face)? What emergency measures would be necessary if this occurred?*

Explore the **submandibular salivary gland** (**6.4.9**) which secretes a seromucous saliva into the mouth. The superficial part of this gland lies on mylohyoid, largely under the body of the mandible, while its deep part tucks around the free posterior edge of mylohyoid (**6.4.6b, 6.4.10**). The superficial part of the gland often protrudes beneath the mandible and can be felt there, otherwise the gland is palpable between an index finger in the floor of the mouth and a thumb pushing upward just anterior to the angle of the mandible.

On a prosection in which mylohyoid has been reflected (**6.4.6b**), identify the deep part of the gland from which the **submandibular duct** emerges to run forward between mylohyoid and hyoglossus muscles before piercing the mucous membrane of the floor of the mouth to open on to the sublingual papilla lateral to the frenulum. The **sublingual gland** overlies the anterior part of the submandibular duct and secretes seromucous saliva which reaches the mouth through a number of small ducts which pierce the mucous membrane of the sublingual ridge.

Blood and lymphatic supply to the tongue and floor of mouth

Locate the **lingual artery** (**6.4.6c**) which arises from the external carotid artery at the level of the hyoid bone and forms an upwardly-directed loop which enters the tongue by passing deep to the posterior border of hyoglossus and passes through the tongue to its tip. It sends dorsal branches upward to supply the tongue, palatoglossal arch, tonsil, and epiglottis, and branches to the salivary glands and gums in the floor of the mouth. These branches anastomose with arteries to the lower jaw. Veins from superficial parts of the tongue form a **lingual vein** which runs with the artery, and drains into the internal jugular vein; those from the tip of the tongue run back on its under-surface and join sublingual veins which pass backward to end variably in the facial, lingual, or internal jugular veins.

Qu. 4I *When the lower jaw is fractured, bleeding is often difficult to stem; why?*

Lymph from the tongue drains by three routes. From the tip of the tongue it passes to submental nodes and thence, via the anterior jugular chain, to the lower deep cervical nodes. From

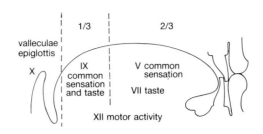

6.4.11
Nerve supply to tongue and epiglottis.

the remainder of the anterior part of the tongue it drains to the submandibular nodes and thence to the upper deep cervical nodes. From the posterior part of the tongue it drains direct to the upper deep cervical nodes. There are extensive anastomoses between these lymphatic routes, especially posteriorly, so that lymph can reach nodes on either side.

Nerve supply to the tongue and floor of the mouth (6.4.6c, 6.4.11)

On the superficial surface of hyoglossus locate the **lingual nerve** (mandibular (V) nerve) which crosses first superficially and then deep to the submandibular duct before supplying the anterior two-thirds of the tongue. At this point, the lingual nerve carries fibres of the mandibular (V) nerve which supply common sensation, and also fibres derived from the facial (VII) nerve which supply sensory fibres to the taste buds (but not the vallate papillae) and secretomotor fibres to the submandibular and sublingual salivary glands (see below). Both common sensation and taste to the posterior one-third of the tongue and to the vallate papillae is derived from the **glossopharyngeal nerve** (IX) which reaches the tongue by passing deep to the hyoglossus. Find this nerve in the neck as it lies close to the lower border of styloglossus. The motor supply to all the tongue muscles (except palato-

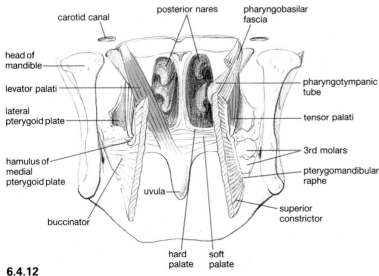

6.4.12
Muscles of palate.

glossus) is from the **hypoglossal nerve** (XII) which can be found on hyoglossus immediately inferior to the submandibular duct.

Locate a small swelling on the lingual nerve as it crosses hyoglossus. This is the **submandibular (parasympathetic) ganglion** which receives preganglionic fibres from the facial nerve via the chorda tympani, and from which pass fine nerves carrying postganglionic parasympathetic fibres which are secretomotor to the submandibular and sublingual salivary glands and isolated glands in the floor of the mouth.

Roof of the mouth — the palate (6.4.12)

Examine a skull and identify the **hard palate** which is formed of the palatine processes of the maxillary bone and the horizontal plates of the palatine bones. Examine the palate on a prosection and define the boundary between hard and soft palate. The hard palate is covered by periosteum and mucous membrane the surface of which is, in part, corrugated; the **soft palate** is a movable fold of mucous membrane which encloses the tendinous aponeurosis of tensor palati, fibres of the palatal muscles, and lymphatic tissue. Identify the mucous membrane-covered 'pillars of the fauces' and the muscles within them (palatoglossus and palatopharyngeus) which link the palate respectively to the side of the tongue and the side wall of the pharynx. At the centre of the palatal arch identify the pendulous, muscular uvula and verify that, when the mouth is closed, it hangs down behind the pharyngeal part of the tongue. On a prosection which exposes the muscles of the palate, identify **tensor palati** which arises outside the pharynx from the outer aspect of the medial pterygoid plate and scaphoid fossa at the root of the pterygoid plates; its tendon passes downward and hooks around the pterygoid hamulus before expanding horizontally to form the fibrous aponeurosis of the soft palate which is attached to the posterior border of the hard palate. **Levator palati** arises within the pharynx from the under-surface of the petrous temporal bone and the adjacent cartilaginous part of the pharyngotympanic tube and passes downward and medially to enter the upper surface of the soft palate. Both muscles act as their names suggest and, with a muscular ridge of the superior constrictor muscle, close off the nasopharynx from the oropharynx during swallowing (p. 68). **Palatoglossus** and **palatopharyngeus** take origin largely from the soft palate and pass downward anterior and posterior to the tonsil to merge with muscles of the side of the tongue and pharynx respectively. They close the oropharyngeal isthmus while food is being chewed and prevent its passing into the oropharynx.

The **arterial** supply to the hard and soft palates is derived from **palatine branches** of the **facial**, **maxillary**, and **ascending pharyngeal arteries**. Its lymph drains into lymphatic tissue around the oropharyngeal isthmus (p. 110) and from thence to upper deep cervical lymph nodes. The **sensory nerves** to the hard and soft palate are derived from the greater and lesser **palatine** and the **long sphenopalatine** branches of the maxillary nerve (V) that are distributed with branches of the pterygopalatine ganglion (p. 120); the soft palate also receives innervation from the **glossopharyngeal** nerve. The **motor** supply derives from the **pharyngeal plexus** except for that of tensor palati which is supplied from the **mandibular nerve** (V).

Parotid gland (6.4.9, 6.4.10)

On a prosection locate the large **parotid** salivary gland which fills the wedge-shaped space between the ramus of the mandible anteriorly and the mastoid process and sternomastoid posteriorly. Its superficial part overlaps the angle of the mandible, and from it the **parotid duct** passes forward across masseter to pierce buccinator and enter the mouth; its deep part extends into the neck around the termination of the external carotid artery to reach the styloid process and its attached fascia (**6.4.10**). It extends upward to the external auditory meatus and temporomandibular joint. The gland is tightly enclosed in fascia which is derived largely from the investing layer of deep cervical fascia; a thickening of this fascia, the stylomandibular 'ligament', separates the parotid from the submandibular gland. Terminal branches of the facial nerve (VII) pass through the gland from its deep posterior aspect, diverge within the gland, and emerge from its anterior border on to the face (p. 123).

The external carotid artery gives branches to the parotid and divides within the gland into maxillary and superficial temporal arteries which emerge, respectively, from the anterior and superior aspects of the gland; likewise the corresponding veins from the retromandibular vein within the gland. Lymph from the parotid drains into the upper deep cervical nodes via superficial glands on the surface of the parotid.

Secretomotor supply to the parotid is derived largely from the **glossopharyngeal nerve** (IX). Preganglionic fibres of the glossopharyngeal nerve synapse in the **otic (parasympathetic) ganglion** (p. 00) from which postganglionic secretomotor fibres travel to the gland with the auriculotemporal nerve. **Sympathetic** fibres also reach the glandular cells and blood vessels. The capsule receives sensory innervation from trigeminal nerve fibres supplying the overlying skin. Locate the auriculotemporal nerve (mandibular division of trigeminal (V) nerve) as it emerges from the posterior part of the gland to supply the ear and scalp. Locate also the facial (VII) nerve which emerges from the skull through the stylomastoid foramen and passes almost immediately into the parotid gland within which it divides into numerous branches which supply muscles of facial expression.

The parotid gland (like other salivary glands) may be subject to a viral infection (mumps) which causes the gland to swell and can be very painful, especially when foods which stimulate the production of saliva are taken.

Qu. 4J *Why should an infected parotid gland be painful?*

Qu. 4K *What structures might be damaged during surgery to remove a tumour of the parotid gland?*

Teeth (6.4.13, 6.4.14)

During the development of the jaws, Man has two sets of teeth: an early deciduous or 'milk' set (without premolars) and a permanent (adult) set. The time of eruption of these teeth varies, but usually conforms to the following scenario.

Deciduous teeth	Erupt at
Central incisors	6–8 months
Lateral incisors	8–10 months
First molars	12–16 months
Canines	16–20 months
Second molars	20–30 months

Permanent teeth	Erupt at
First molars	6–7 years
Central incisors	6–8 years
Lateral incisors	7–9 years
Canines	9–12 years
First/second premolars	10–12 years
Third molars ('wisdom')	17–21 years

Examine the teeth; each consists of a **crown** covered by hard translucent **enamel**, and a **root** which is covered with **cement**. The crown and the root meet at the **neck**. Within lies the **dentine**, inside which again is a **pulp cavity** which extends from the root to the crown. The pulp canal transmits through its **apical foramen** a

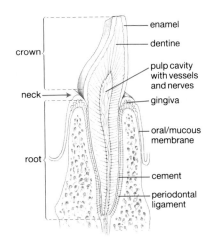

6.4.13
Sagittal section through incisor tooth.

vascular and a sensory nerve supply to the dentine. Each tooth has a characteristic function-related morphology (**6.4.13**). The incisors have biting edges and are concave lingually; they have single roots. Canines have a single cusp and a very long root which can be palpated through the face. Premolars have two cusps and one or two roots. The first and second molars have four or more cusps and two or three roots. Third molar teeth (wisdom teeth) are smaller, and may fail to erupt or be absent (**6.4.14**).

Each tooth is held in a bony socket in the jaw by a 'peg and socket' fibrous joint (gomphosis). The cement of the tooth is held to the alveolar bone by the **periodontal ligament** which holds the tooth firmly in place, but allows a little movement during biting and chewing, and permits the tooth to erupt. The periodontal ligament is continuous with the subcutaneous tissue of the gums.

Qu. 4L *What is the function of incisor, canine, premolar, and molar teeth?*

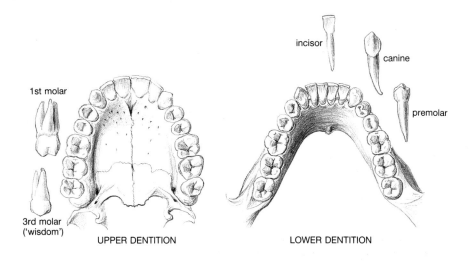

6.4.14
Teeth of upper and lower jaws.

C. Radiology

Examine the CT which shows a transverse section through the parotid gland (**6.4.15**).

Examine the sialogram of the submandibular gland (**6.4.16**). Note the point at which the tip of the catheter containing contrast medium has entered the orifice of the duct in the floor of the mouth (arrow). This gland is more compact than the parotid gland and the intraglandular ducts are often shorter and wider than those of the parotid.

Examine the **sialogram** of the parotid gland in lateral view (**6.4.17a**). Contrast medium has been injected through a catheter which can be seen entering the duct opposite the second upper molar tooth (arrow). The main parotid duct can be traced posteriorly (superficial to masse-

ter) and its fine tributaries within the substance of the gland are visible over the posterior margin of the ramus of the mandible. The first tributary on its superior aspect drains the accessory part of the parotid gland, which is situated above the main duct (arrow head). The frontal view (**6.4.17b**) shows the catheter entering the duct within the mouth (arrow). Intraglandular ducts can be seen in the deep part of the gland, medial to the posterior margin of the mandible.

Requirements:

Skull and mandible.

Prosections of the floor of the mouth and tongue, from the lateral aspect, at the level of mylohyoid, hyoglossus, and deep to hyoglossus; a hemisected head; prosection of the submandibular gland, parotid gland *in situ* and parotid bed.

Sialograms, CTs through mouth.

Examination gloves, disposable tongue depressor, tissues, mirror, torch.

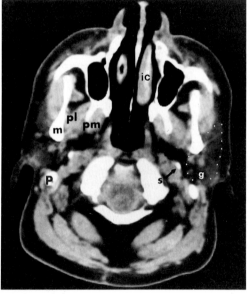

6.4.15
Horizontal CT through parotid gland (dotted outline); deep part of gland (g). Neck of mandible (m); lateral (pl) and medial (pm) pterygoid muscles; mastoid process (p); styloid process (s).

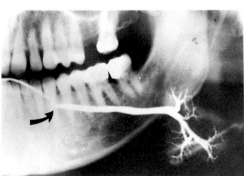

6.4.16
Sialogram of submandibular gland.

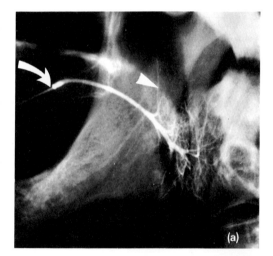

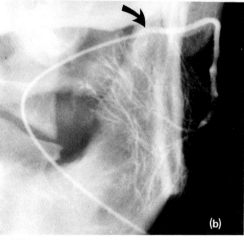

6.4.17
Sialogram of parotid gland: (a) lateral aspect; (b) frontal aspect.

Seminar 5

Pharynx and oesophagus

The **aim** of the seminar is to study the pharynx and that part of the oesophagus which lies within the neck, and their movements during swallowing (deglutition).

Development of pharynx and oesophagus

As the primitive heart moves ventrally and caudally from its initial rostral position, the foregut is formed as an endodermal diverticulum dorsal to it (see **5.2f**). Rupture of the oropharyngeal membrane brings the endodermal foregut into continuity with the ectodermal stomodeum, forming a primitive oral cavity at the rostral end of the embryonic pharynx.

The first aortic arch forms within the mesenchyme in the lateral wall of the oral cavity; it joins the truncoconal region of the heart to the paired dorsal aortae. The mesenchyme around the aortic arch is at first derived from paraxial mesoderm but a mass immigration of neural crest-derived mesenchymal cells results in the formation of mandibular and maxillary prominences which comprise the first pharyngeal arch. The 2nd and subsequent arches are formed in a similar manner: the heart continues to move caudally, aortic arch arteries form, and neural crest cells migrate into the mesenchyme around them. Five pharyngeal arches form on each side (numbered 1, 2, 3, 4, and 6, because, by analogy with lower vertebrates, mammals have lost the 5th arch).

Between adjacent pharyngeal arches there is a mesenchyme-free region where surface ectoderm and pharyngeal endoderm are in contact, to form a **pharyngeal groove** on the outer surface and a **pharyngeal pouch** on the inner surface. The dorsal part of the 1st pouch (with contributions from the 2nd and 3rd pouches) elongates to form the primitive middle ear cavity and pharyngotympanic tube which connects the middle ear with the pharynx. Derivatives of the other pouches are summarized in Chapter 5.

During development of the face, formation of the secondary palate divides the primitive oral cavity into definitive oral and nasal cavities which open separately into the foregut, defining the nasopharynx and oropharynx. The developing larynx opens into the ventral aspect of the most caudal part of the pharynx to define the laryngopharynx. Motor and sensory innervation of the adult pharynx is derived from the nerves of the 3rd and 4th arches, i.e. the glossopharyngeal (IX) and vagus (X) nerves.

A. Living anatomy

The **pharynx** is a muscular half-tube which lies in front of the upper six cervical vertebrae and which is continuous with the oesophagus. It is described in three parts: a nasal part — the **nasopharynx** — which is continuous anteriorly with the internal nares and connected on each side with the middle ear by the pharyngotympanic (Eustachian) tube; an oral part — the **oropharynx** — which is continuous anteriorly with the mouth; and a laryngeal part — the **laryngopharynx** — which lies behind the larynx and the aditus to the larynx and trachea. At its lower extremity, the laryngopharynx opens into the oesophagus, a muscular tube which conveys swallowed material through the neck and thorax to the stomach.

Ask your partner to open his mouth widely and protrude his tongue. Depress the tongue with a disposable spatula and ask your partner to say 'aah'. The resultant elevation of the soft palate (p. 56) should enable you to see the pillars of the fauces and the tonsillar fossa on either side of the soft palate and uvula, behind which lies the posterior wall of the oropharynx.

To see into the nasopharynx or laryngopharynx it is necessary to use an angled mirror inserted gently through the mouth and oropharyngeal isthmus and directed either upward or downward; the procedure requires skill so ask a member of staff familiar with the procedure to show you how to do it.

Blow out through your mouth as if you were blowing a trumpet. Note that no air escapes through the nose.

Qu. 5A   *What is happening inside the pharynx? (It also happens when explosive consonants such as d, p, or b are spoken).*

With your index finger and thumb on either side of the neck immediately beneath the lower jaw, gently feel the hyoid bone and its greater horns and, below it, the thyroid and cricoid laryngeal cartilages (p. 72). Ask your partner to take a mouthful of water and watch the movement of the hyoid and larynx as the water is swallowed. Palpate the greater horns of the hyoid and feel the movement that occurs. Note also the movements of the tongue when you swallow.

Qu. 5B   *What movement does the hyoid and larynx make during swallowing?*

Qu. 5C   *What happens to the tip of your tongue when you swallow?*

Now try to swallow a little water first with your mouth shut, and then with your mouth open.

Qu. 5D *Which is the more difficult and why?*

B. Prosections of the pharynx

Interior of the pharynx (6.5.1; 6.5.2)

Examine a prosection in which the interior of the pharynx is exposed from its posterior aspect, together with a midline hemisection (**6.5.2**).

6.5.1
Interior of pharynx, posterior aspect.

The **nasopharynx** is the uppermost part of the pharynx and is continuous below with the oropharynx and anteriorly with the nose. It has a conductive respiratory function; therefore, its mucous membrane is lined with ciliated epithelia. Examine its roof, which is formed by the basisphenoid and basiocciput, and its posterior and lateral walls which are formed largely by the superior constrictor muscle and its associated fascia (see below). Beneath the mucous membrane and the bone is lymphoid tissue (**adenoids**) which are naturally larger in children than adults and can become very enlarged in children below the age of 7 years; enlarged adenoids can hang from the roof of the nasopharynx immediately behind the internal nares thereby obstructing the airway and causing the child to breathe through an open mouth.

Locate the opening of the **pharyngotympanic (Eustachian) tube** immediately posterolateral to the internal nares and note its relation to the middle and inferior concha of the nose. As its name indicates, the pharyngotympanic tube connects the nasopharynx to the middle ear. The upper, lateral third of the tube is bony. On a skull, locate the medial end of the bony part of the tube at the junction of the petrous and squamous parts of the temporal bone (i.e. posteromedial to the foramen spinosum); pass a fine probe through it until the tip of the probe in the middle ear can be seen through the external auditory meatus. The lower, medial two-thirds of the tube is triangular in section and its pharyngeal end raises the **tubal elevation**. The anterior and posterior walls are cartilaginous, the floor fibrous. From its posterior wall, a fold of mucous membrane covering the salpingopharyngeus muscle passes downward while im-

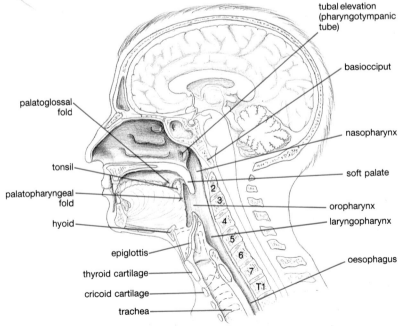

6.5.2
Sagittal section of head and neck to show pharynx.

mediately below the tube levator palati passes down into the soft palate.

Posterolateral to the opening of each pharyngotympanic tube locate the deep **lateral recesses** of the nasopharynx in which nasopharyngeal tumours can grow undetected for a while (**6.5.5**).

Qu. 5E *If you wished to pass a cannula through the nose and into the pharyngotympanic tube, under which concha would you pass it?*

Qu. 5F *Which bones form the roof of the nasopharynx and the posterior aspect of the septum of the nose?*

The **oropharynx** is continuous with the nasopharynx above and the laryngeal part of the pharynx below. Its lateral and posterior walls are formed by the superior and middle constrictor muscles (see below). Anteriorly is the oropharyngeal isthmus, the posterior third of the tongue, and upper part of the epiglottis.

Locate the palatoglossal and palatopharyngeal folds with their underlying muscles and any lymphatic tissue of the **palatine tonsil** (characterized by pits which pass into the tonsillar **crypts**) visible between them. The collections of submucosal lymphoid tissue which lie in the walls of the oro- and nasopharynx form a ring around the upper respiratory tract.

Qu. 5G *What is the function of this lymphoid tissue?*

The **laryngopharynx** is continuous with the oropharynx above and the oesophagus below. Its lateral and posterior walls are formed largely by fibres of the inferior constrictor muscle. Both the oro- and laryngopharynx come into contact with food; thus their lining epithelium is of the stratified squamous type. The epiglottis, larynx, and pyriform fossae form its anterior wall. Examine the **epiglottis** and, on a midline section, note that this (elastic) cartilage passes upward and backward from the thyroid cartilage, behind the hyoid bone and tongue. Depress the epiglottis over the laryngeal opening and note its connection to the base of the tongue by a median **glosso-epiglottic fold** on either side of which are small pouches, the **valleculae**, bounded by lateral folds. The **aditus** to the **vestibule** of the larynx is protected by the epiglottis, the upper part of which acts as a roof over the aditus and diverts swallowed material laterally into the pyriform fossae (see below).

Examine the margins of the aditus to the larynx. Anteriorly is the epiglottis and, from its sides, two **aryepiglottic folds** consisting of muscle covered by mucous membrane, pass down to attach to the **arytenoid cartilages** of the larynx (p. 72). The aditus is completed by muscles and mucosa between the two arytenoid cartilages. Within the aryepiglottic folds palpate two pairs of small cartilaginous nodules (corniculate and cuneiform) which give added support to the wall of the laryngeal vestibule. Lateral to the laryngeal vestibule and aryepiglottic folds are **piriform fossae** which act as channels on either

side of the laryngeal opening through which swallowed food and liquids pass to the oesophagus. Each pyriform fossa is bounded laterally by the thyrohyoid membrane and the inner surfaces of the laminae of the thyroid cartilage (p. 72); if food becomes lodged in this position, a bout of reflex coughing may be set up which is difficult to control until the foreign body is dislodged. The lowest part of the laryngopharynx lies behind the cricoid cartilage (p. 72) and is continuous with the oesophagus at the level of the 6th cervical vertebra.

Muscular walls of the pharynx
(**6.5.3; 6.5.4**)

Examine a pharynx dissected from the exterior. The pharynx is essentially a muscular half-tube, which is attached above to the base of the skull and, anteriorly, to the posterior margins of the nose, mouth, and larynx. Between the mucous membrane and muscle layers is a thickened fibrous layer which is most prominent beneath the skull where it forms the **pharyngobasilar fascia**. Three muscles form the half-tube, the **superior, middle,** and **inferior pharyngeal constrictors**. Blending with these are pairs of vertically-disposed muscles (stylopharyngeus, palatopharyngeus, and salpingopharyngeus). On a hemisected head (**6.5.2**) note that the pharynx lies in front of the upper six cervical vertebrae and the prevertebral muscles and prevertebral

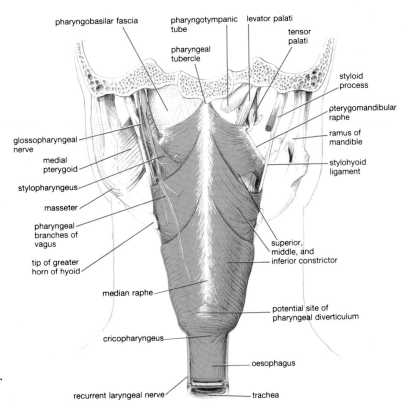

6.5.3
Exterior of pharynx, posterior aspect.

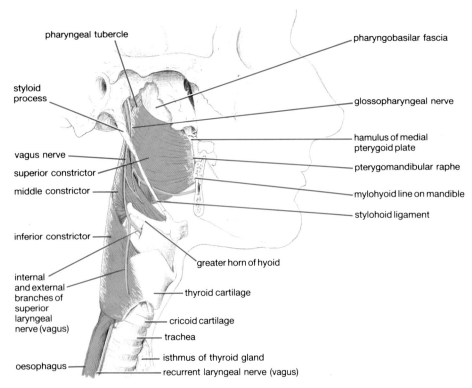

6.5.4
Pharynx, lateral aspect.

Above the free edge of the superior constrictor feel for the cartilaginous inferior part of the pharyngotympanic tube and, on the skull, locate where the tube pierces the pharyngobasilar fascia.

Revise the attachments of tensor and levator palati (p. 59) in relation to the superior constrictor. Both muscles are attached to the pharyngotympanic tube, but levator palati lies entirely within the nasopharynx, whereas tensor palati lies outside the pharynx until its tendon hooks round the hamulus. During swallowing (and sometimes during speaking) the soft palate is tensed and raised by tensor and levator palati (see **6.4.12, 6.5.1**) against a ridge on the posterior wall of the pharynx (Passavant's ridge), thus closing off the nasopharynx from the oropharynx. The ridge is produced by contraction of fibres of superior constrictor which lie at the junction of the naso- and oropharynx.

Examine the **middle constrictor**. It arises from the angle between the stylohyoid ligament and the greater cornu of the hyoid bone. Its upper fibres overlap externally the lower fibres of the superior constrictor and some may even reach the pharyngeal tubercle. Between the upper margin of middle constrictor and the lower margin of superior constrictor, stylopharyngeus (see below), the lingual artery, glossopharyngeal nerve (IX), and pharyngeal branches of the vagus nerve (X) gain access to the interior of the pharynx and tongue.

The **inferior constrictor** consists of two parts which arise respectively from an oblique line on the side of the thyroid cartilage and from the side of the cricoid cartilage which lies immediately below. Trace the muscle fibres backward and note that the upper fibres (thyropharyngeus) again overlap those of the muscle above and form a posterior raphe. Fibres originating from the cricoid (**cricopharyngeus**), however, pass around the pharynx as a U-shaped sphincteric loop and merge below with fibres of the oesophagus. These fibres maintain a sphincteric tone until swallowing begins, when

fascia that cover their anterior aspect. Between the pharynx and the prevertebral fascia is a potential space — the **retropharyngeal space** — which extends down into the posterior mediastinum; this enables the pharynx to move easily on the vertebral column during swallowing.

Qu. 5H *Occasionally, sharp foreign bodies are swallowed and become lodged in the laryngopharynx. If they are not quickly removed they may perforate the wall and infection will ensue. What is the danger if such a condition occurs?*

Examine the **superior constrictor** which arises from the pterygomandibular raphe and the adjoining bone at either end — the hamulus of the medial pterygoid plate and the mandible behind the last molar tooth.

Qu. 5I *What other muscle is attached to the pterygomandibular raphe?*

Note that the fibres of superior constrictor spread backward, fanwise, to meet those of the opposite side in a median raphe. The raphe and uppermost fibres extend upward to the **pharyngeal tubercle** on the basiocciput; note the position of this tubercle on a skull. Next trace the uppermost muscular fibres of the superior constrictor which form a free edge between the medial pterygoid plate and pharyngeal tubercle; above this the wall of the nasopharynx is formed by the strong **pharyngobasilar fascia**. Trace the attachments of the pharyngobasilar fascia to the base of the skull (**6.5.5**) and note the position of the lateral recess of the nasopharynx.

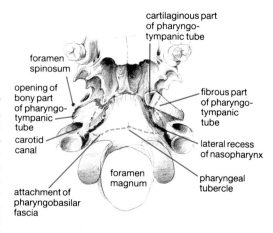

6.5.5
Attachment of pharyngobasilar fascia to base of skull.

they relax; after swallowing, sphincteric tone prevents regurgitation of food into the laryngopharynx. The sphincteric action may become uncoordinated. If this happens the pressure generated by the constrictors during swallowing eventually can cause the formation of an out-pouching (pharyngeal diverticulum) at the weak point between the posterior raphe and the sphincter (Killian's dehiscence) (6.5.3). Between the upper fibres of inferior constrictor and the lower fibres of middle constrictor the internal laryngeal nerve and vessels pass through the thyrohyoid membrane to supply the larynx. At the lower border of inferior constrictor the recurrent laryngeal nerve and the inferior laryngeal artery gain access to the larynx by passing upward between the muscle and the larynx (p. 175).

Identify three muscles which merge with the pharyngeal constrictors. On the outside of the pharynx locate **stylopharyngeus** which passes downward from the styloid process to blend with the inner aspect of the middle constrictor; within the pharynx locate **palatopharyngeus** beneath the palatopharyngeal fold and **salpingopharyngeus** (a few muscle fibres which run downward from the pharyngotympanic tube to blend with palatopharyngeus).

Qu. 5J *When and why might you wish voluntarily to contract salpingopharyngeus?*

Examine the **oesophagus** which is continuous with the cricopharyngeal sphincter at the level of the cricoid cartilage. It has only a short course in the neck, where it lies anterior to the 6th and 7th cervical vertebra and posterior to the trachea, before it passes into the posterior mediastinum of the thorax and thence to the stomach (Vol. 2, Chapter 5, Seminar 5). It is narrower than the pharynx, and lined by a non-keratinized, stratified squamous epithelium pierced by small oesophageal mucous glands. In its upper part the outer longitudinal and inner circular coats are formed from striated muscle; but in the lower third, as the stomach is approached, this changes to smooth muscle. In the neck the oesophagus passes from the midline slightly to the left. On either side are the common carotid artery, the lateral lobes of the thyroid gland, the sympathetic trunks, and the recurrent laryngeal nerves which lie in the groove between trachea and oesophagus. Locate these structures which you will study in later seminars.

Qu. 5K *One further structure, which you may already have encountered in the thorax, lies close to the oesophagus in the neck, but only on the left side. What is it and what is its function?*

Blood, lymphatic, and nerve supply to the pharynx and oesophagus
(6.5.6)

The pharynx receives arterial blood from a number of arteries: the **ascending pharyngeal** (a

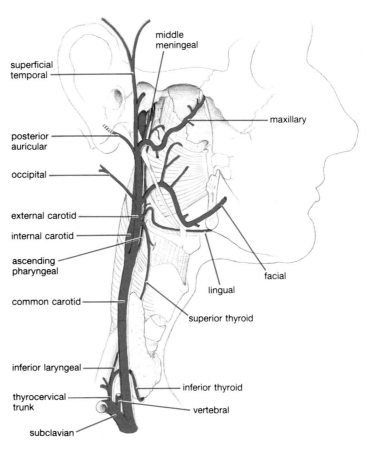

middle meningeal

superficial temporal

maxillary

posterior auricular

occipital

external carotid

internal carotid

facial

ascending pharyngeal

lingual

common carotid

superior thyroid

inferior laryngeal

inferior thyroid

thyrocervical trunk

vertebral

subclavian

6.5.6
Arterial supply to pharynx.

branch of the external carotid at its formation); the **inferior** and **superior thyroid**; the **lingual** and **facial arteries**. Where it lies lateral to the tonsillar fossa, the loop of the facial artery gives a branch to the palatine tonsil; this is at risk in tonsillar surgery. Vessels to the cervical part of the oesophagus are derived largely from the inferior thyroid artery and veins. Venous blood drains into a **pharyngeal plexus of veins** which lies between the constrictor muscles and the prevertebral fascia and which drains into the internal jugular vein. **Lymph** drainage of the pharynx and oesophagus is into the deep cervical nodes either directly or via paratracheal nodes.

The **nerve** supply to the pharynx is derived largely from the **pharyngeal plexus** which receives contributions from the **glossopharyngeal** (IX), **vagus** (X), and **cranial accessory** (XI) nerves, and from autonomic fibres. The nerve supply to the oesophagus in the neck is derived from the **recurrent laryngeal** branch of the vagus nerve and cervical **sympathetic** trunk. (For the supply of the thoracic and abdominal parts of the oesophagus see Vol. 2, Chapter 5, Seminar 5.)

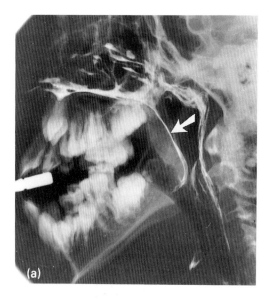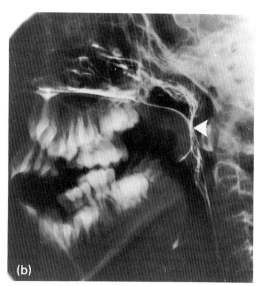

6.5.7
Radiographs showing position of soft palate: (a) at rest; (b) during speech.

C. Radiology

Movements of the soft palate

Examine **6.5.7a** and **b** which are radiographs of the mouth and palate of a child (note the primary and secondary dentition). A small amount of a barium-containing compound has been sprayed into the nose so that the walls of the nasopharynx and the dorsal surface of the hard and soft palates are clearly seen.

In the radiograph taken when the child was breathing normally (**6.5.7a**) the palate (arrow) is in a neutral position and both nasal and oral airways are open. Compare **6.5.7b** taken when the child was speaking: the palate has been tensed and elevated so that it reaches to the posterior pharyngeal wall and closes the nasopharynx. To assist closure, the constrictor muscles of the nasopharynx contract slightly, raising a prominence on the posterior wall of the nasopharynx (Passavant's ridge, arrow head). The posterior portion of the palate consisting of the uvula has no part in palatal closure.

Movements of the pharynx during swallowing (6.5.8a–i)

In this series of radiographs a suspension of barium sulphate, which appears black on the radiographs, is being swallowed. The frames are selected from a cineradiography study of the swallowing cycle.

(a) When barium (B) is taken into the mouth, the posterior part of the tongue is elevated against the palatoglossal arches so that the fluid is held in the mouth in a midline groove on the tongue. Breathing is possible through the nasopharyngeal airway (A). Note the position of the hyoid bone (H) and epiglottis (arrow).

(b) Swallowing is started voluntarily: the tip of the tongue is elevated against the hard pa-

late and the teeth, the palatoglossal folds are relaxed, and the posterior part of the tongue is depressed, so that barium is transferred on to the posterior one-third of the tongue. As it passes the fauces the swallowing reflex is triggered through the glossopharyngeal nerve.

(c) As the pharyngeal phase of swallowing begins, the palate is tensed and elevated, thereby closing the nasopharyngeal airway to prevent reflux of fluid into the nose. Expulsion of the material from the mouth into the oropharynx is accomplished by upward and backward movement of the tongue, which occurs in part by elevation of the floor of the mouth.

(d) Barium is propelled through the pharynx by peristalsis as the contrictors first relax to accommodate the bolus and then contract behind it. The epiglottis inverts over the laryngeal vestibule as the hyoid bone is elevated, raising the laryngeal skeleton beneath the tongue. This is best seen over the course of the next three frames. At the same time, the aditus to the larynx is narrowed and the glottis is closed

Qu. 5L *Which muscles and their nerve supply are primarily involved in the movements (a)–(d)?*

(e) As the bolus reaches the distal pharynx the circular fibres of the inferior pharyngeal constrictor muscle (cricopharyngeal sphincter) relax and barium begins to pass into the upper oesophagus.

(f) As the bolus passes, the pressure in the pharynx is at its greatest and the cricopharyngeal sphincter is widely open.

(g) As the pharynx empties, the pharyngeal pressure falls and the cricopharyngeal sphincter begins to close.

(h) The pharynx is empty and cricopharyngeus contracted, closing the oesophagus. The ele-

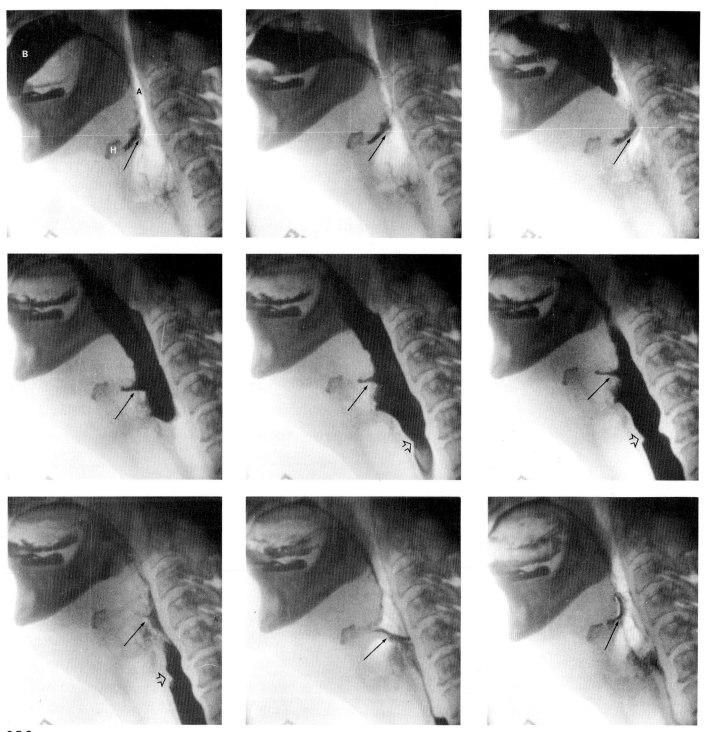

6.5.8
Radiographs showing progressive stages of swallowing (a)–(i).

vators of the hyoid begin to relax and the larynx starts to return to its normal position. The epiglottis is still inverted.

(i) Finally, the epiglottis flicks back into its original position (arrow) and swallowing may begin again. The entire cycle takes no more than 1–2 seconds.

For reflex control of swallowing see p. 142.

Requirements:
Articulated skeleton, separate skull.
Prosections of pharynx from lateral and posterior aspects; hemisected head and neck; pharynx opened posteriorly; blood and nerve supply of pharynx and oesophagus.
Radiographs illustrating swallowing.
Disposable tongue depressors, disposable gloves.

Seminar 6

Larynx, trachea, and thyroid gland

The **aim** of this seminar is to study the larynx and trachea in the neck, and their function in phonation and respiration; also to study the thyroid gland.

Development of the larynx and trachea

During the fourth week of fetal life a **laryngotracheal groove** appears on the ventral aspect of the lower end of the embryonic pharynx in the region of the 4th and 6th pharyngeal arches (see **6.4.1**). The endoderm of the groove forms the epithelium and glands of the larynx and trachea; the surrounding mesoderm forms the connective tissue, muscle, and cartilage. As the groove elongates and deepens two **tracheo-oesophageal folds** appear on either side. These increase in size, grow toward each other, and fuse in the midline, thus separating the developing trachea from the more dorsal oesophagus. The tracheal diverticulum enlarges caudally and divides repeatedly to form the lower respiratory tract (Vol. 2, Chapter 5, Seminar 5). The oesophagus elongates and narrows so that its lumen becomes occluded and must later recanalize (Vol. 2, Chapter 5, Seminar 5). Abnormally, this recanalization can be incomplete (oesophageal atresia or stenosis) or the connection between the oesophagus and trachea may close incompletely or reopen (tracheo-oesophageal fistula) and the two problems may coexist. If the oesophagus fails to recanalize, the fetus cannot swallow amniotic fluid which therefore accumulates (polyhydramnios). Any abnormal connection between the trachea and oesophagus, or oesophageal constriction, can allow milk swallowed by the newborn child to enter the airway, which can then become obstructed.

The cranial aspect of the laryngotracheal diverticulum remains connected to the pharynx to form the **aditus to the larynx**. Around this, the 4th and 6th arch tissues form the **cartilages** and **muscles of the larynx** which are supplied by the 4th and 6th arch nerves (external and recurrent laryngeal branches of the vagus). Within the larynx the vocal folds mark the division between 4th and 6th arch endoderm; this division is reflected by the innervation of the mucosa (internal laryngeal branch of the vagus (4th arch) above, and recurrent laryngeal branch of the vagus (6th arch) below the vocal folds).

A. Living anatomy (6.6.1)

Palpate the **hyoid bone** (H) and identify its midline **body**, and the **greater horns** which extend backward at C3–C4 level. Now look at the anterior aspect of the neck of people around you and locate the midline laryngeal prominence (LP) ('Adam's apple') below the hyoid (**6.6.1**).

Qu. 6A *Is the laryngeal prominence larger in males or females?*

The prominence is caused by the anterior aspect of the **thyroid cartilage** where the two **laminae** which comprise it are joined. Explore the extent of the laminae on yourself and your partner by grasping the structure between your forefinger and thumb; trace the upper border of the cartilage, and try to feel the upward projecting **superior horns** at its posterior extremity. Feel for the gap between the greater horn of the hyoid and the upper border of the thyroid; this is occupied by the **thyrohyoid membrane**. Now feel down the midline of the thyroid cartilage to its lower limit. Immediately beneath the thyroid cartilage feel the very narrow depression occupied by the **cricothyroid ligament** and, below that, the narrow anterior arch of the **cricoid cartilage** which lies at the C6 level. Below the cricoid cartilage, feel the individual cartilaginous

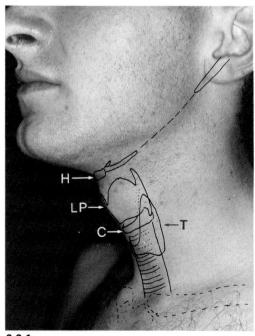

6.6.1
Laryngeal prominence and surface markings of laryngeal cartilages.

'rings' of the **trachea** as it passes in the midline down to the suprasternal notch. Count the number of 'rings'; the 2nd and 3rd are difficult to feel because the isthmus of the thyroid gland lies over them.

To visualize the interior of the larynx (**6.6.2**) it is necessary to insert a small angled mirror through the open mouth and into the oropharynx. A source of light is needed; the tongue should be gently depressed; and the soft palate elevated by the subject saying 'aah'. To gain a good view requires considerable expertise. **6.6.2** shows the appearance of the living larynx. Note the **aryepiglottic folds** which form the lateral margins of the laryngeal inlet. Within the larynx identify the **vocal folds** and the narrow slit between them, the **rima glottidis**. The vocal folds often appear more pale in colour than the surrounding mucosa because of the attachment of their stratified squamous epithelium to the underlying tissue. The remainder of the larynx and trachea is lined by pseudostratified ciliated respiratory epithelium.

Qu. 6B *For what reason are the vocal folds covered with stratified squamous epithelium? Where else in this region would you expect to find such epithelium?*

Above the vocal folds, the **vestibular folds** do not reach so close to the midline. During quiet respiration there should be a distinct gap between the vocal cords; on forced breathing the gap is enlarged; any attempt to make vocal sound causes the vocal folds to come together and to vibrate. In all situations, the rima and the movements of the two vocal folds should be symmetrical.

Palpate again the external aspect of the larynx and of the trachea, and swallow.

Qu. 6C *How does the larynx move when you swallow?*

Finally, feel down the lateral aspect of the thyroid cartilage laminae toward the cricoid. You may just be able to feel the soft tissue that comprises the lateral lobes of the **thyroid gland**. These are united in the midline by a narrow isthmus, usually at the level of the 2nd or 3rd ring of the trachea. The thyroid is slightly larger in females than males, and tends to become enlarged at menstruation and in pregnancy.

B. Prosections of the larynx

The cavity of the larynx and trachea (6.6.3, 6.6.5)

Examine the cavity of the larynx (**6.6.3**). Identify the **inlet to the larynx**, bounded by the epiglottis anteriorly, aryepiglottic folds laterally, and interarytenoid tissues posteriorly. The inlet leads into the **vestibule of the larynx** which extends down to the vestibular folds. Between the vestibular and vocal folds of each side is a small recess, the **sinus of the larynx** (ventricle of larynx), which leads laterally and upward to the saccule of the larynx. (The saccule is small in Man and contains only a few mucous glands which keep the vocal folds moist.) The cavity of the larynx is most narrow at the **glottis** between the vocal folds; beneath the glottis the cavity widens within the cricoid, to become continuous with the trachea. Above the glottis, the laryngeal sinuses together form the **ventricle** of the larynx.

The trachea begins at the lower border of the cricoid cartilage (C6), runs down through the neck in the midline, enters the thorax, and ends behind the 2nd right costal cartilage (T4); it is therefore about 10 cm in length. Note that, although it is roughly tubular in shape, its posterior wall is flattened since the 'rings' of cartilage which maintain its patency are incomplete where it impinges on the oesophagus.

Qu. 6D *What is the functional advantage of this arrangement?*

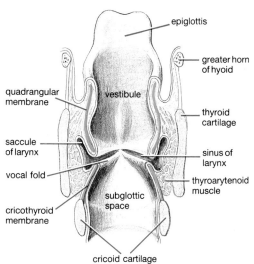

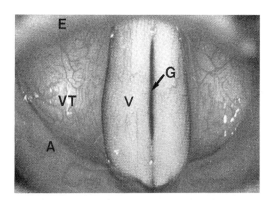

6.6.2
Endoscopic view of vocal folds during phonation. Vocal fold (V); vestibular fold (VT); glottis (G); epiglottis (E); aryepiglottic fold (A).

6.6.3
Interior of larynx, coronal view.

Bones and cartilages of the larynx and trachea (6.6.4, 6.6.5)

Review the structure of the **hyoid bone** (p. 32); its body, greater and lesser cornua (horns), and then examine a prosection dissected to show the bones, cartilages, ligaments, and membranes of the larynx.

Examine the **thyroid cartilage** and identify its two quadrilateral **laminae** which are partly joined in the midline but, above the laryngeal prominence, are separated by the V-shaped thyroid notch. Slender **superior** and **inferior horns** prolong the posterior border of the laminae and, on the lateral surface of the laminae, an **oblique line** curves downward and forward from the base of the superior horn.

The **cricoid** cartilage (**6.6.6**) is shaped like a signet ring and comprises a narrow anterior arch and a large, roughly quadrate posterior lamina. Note that the inferior horn of the thyroid cartilage overlaps and makes a synovial articulation with the posterolateral part of the arch of the cricoid. Note also that the inferior border of the cricoid is horizontal.

Identify the two **arytenoid** cartilages and their various processes (**6.6.7**). These elastic cartilages are shaped like triangular pyramids, the bases of which articulate with the sloping posterolateral aspects of the lamina of the cricoid by synovial joints. Each arytenoid has a pointed **vocal process** which extends anteriorly and to which the vocal fold is attached, a blunt lateral **muscular process** to which certain muscles are attached, and an **apical process** to which is attached the aryepiglottic fold.

Examine the movement possible at the synovial joints between these three cartilages, remembering that the extent of movement is far greater in the living. If the arytenoids are merely rotated on the cricoid, then lateral movement of the vocal process, which widens the glottis, is associated with medial movement of the muscular process. However, during opening of the glottis, the entire arytenoid cartilages are also pulled downward and laterally on the sloping shoulder of the cricoid. Next, gently tilt the

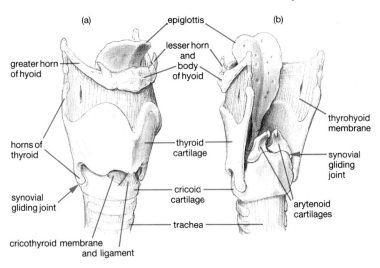

6.6.4
Laryngeal cartilages and membranes. (a) Anterior; (b) posterior.

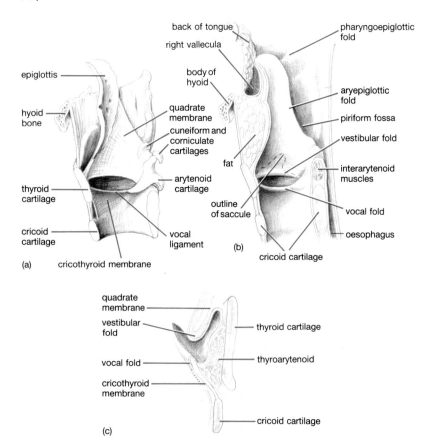

6.6.5
Larynx. (a) Arytenoid cartilage; (b) internal aspect of mucosa; (c) internal aspect of cartilages and membranes.

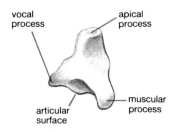

6.6.6
Left arytenoid cartilage.

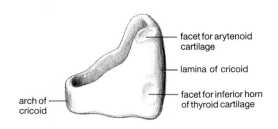

6.6.7
Cricoid cartilage.

thyroid cartilage forwards on the cricoid and note that this movement stretches the vocal fold which extends between the thyroid cartilage and the vocal process of the arytenoid. The vocal fold is also stretched when the thyroid cartilage is pulled forward at the loose synovial joint between the inferior horns of the thyroid and the cricoid.

Examine the **epiglottis** (**6.6.5b**), a thin leaf-shaped lamina of elastic cartilage. Its stem is attached by elastic tissue to the inner aspect of the angle between the thyroid laminae and it extends upward behind the body of the hyoid bone and the back of the tongue. From its lateral margins, the aryepiglottic folds pass downward to the arytenoid cartilages, defining the inlet to the larynx. Small corniculate and cuneiform cartilages lie within and strengthen the folds. The free end of the epiglottis is broad and rather tent-shaped so that, when it approximates to the laryngeal inlet during swallowing, it diverts the food into the **pyriform fossae** on either side of the larynx (**6.6.5a**).

Between the upper border of the thyroid cartilage and the **upper** posterior border of the hyoid bone and its greater horns is attached the **thyrohyoid membrane**. With the inner surface of the lamina of the thyroid cartilage, this membrane forms the lateral wall of the pyriform fossa. Passing from the sides of the epiglottis to the apex and sides of each arytenoid cartilage is a thin **quadrate membrane** of fibroelastic tissue. With its covering mucosa it forms the lateral wall of the vestibule of the larynx, and the medial wall of the pyriform fossa of the pharynx.

Qu. 6E *Which folds do the mucous membrane-covered upper and lower margins of the quadrate membrane form?*

Within the vocal folds and intimately attached to their mucous membrane is the upper, free margin of the well-defined, elastic **cricovocal membrane** (**6.6.5b**). Its base is attached to the upper margin of the cricoid cartilage; its posterior extremity to the vocal process of the arytenoid cartilage; and its anterior extremity to the inner midline angle of the thyroid cartilage. Thus, the glottis consists of an intermembranous anterior portion, and an intercartilaginous portion between the two arytenoid cartilages. In the anterior midline the membrane passing up from the cricoid is thickened to form the **cricothyroid ligament**.

Qu. 6F *If the mucous membrane above the vocal cords began to swell due to a severe throat infection or a bee sting in the throat, what would be the danger?*

Muscles and movements of the larynx

The movements are conveniently considered as: those that protect the laryngeal airway during swallowing; those that occur during variations in respiration; and those that occur during phona-

tion. The muscles involved are: intrinsic muscles of the larynx which move the cartilages relative to one another; muscles of the pharynx that gain attachment to the laryngeal cartilages (p. 65); and the **'strap' muscles** of the front of the neck.

Intrinsic muscles of the larynx (**6.6.8, 6.6.9**)
Identify the **aryepiglottic** muscle fibres which lie within the aryepiglottic folds, and the transverse and oblique **interarytenoid** muscles which unite the two arytenoid cartilages posteriorly (**6.6.8**). Locate two muscles inserted into the muscular process of the arytenoid. Fibres of **posterior cricoarytenoid** arise from the posterior aspect of the cricoid; its fibres pass upward and laterally such that the upper fibres are almost horizontal, the lower almost vertical. **Lateral cricoarytenoid** arises from the upper border of the arch of the cricoid cartilage and pass upward and backward to the muscular process. Fibres of **thyroarytenoid** pass from the inner aspect of the angle between the thyroid laminae backward to the vocal process of each arytenoid; its fibres form a thin muscular sheet that lies lateral to the vocal fold, cricovocal membrane, sinus, and saccule of the larynx. A distinct bundle of its fibres — **vocalis** — lies parallel and just lateral to the vocal ligament, to which some muscle fibres are attached.

Only **cricothyroid** lies on the external aspect of the larynx. Its fibres fan back from the lateral aspect of the cricoid arch to attach to the lower border of the lamina of the thyroid and the anterior aspect of its inferior horn.

All intrinsic muscles of the larynx are supplied by the recurrent laryngeal branch of the vagus nerve except for cricothyroid which is supplied by the external laryngeal branch of the vagus.

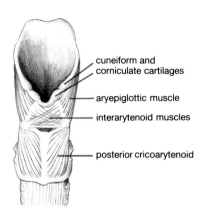

6.6.8
Muscles of larynx, posterior view.

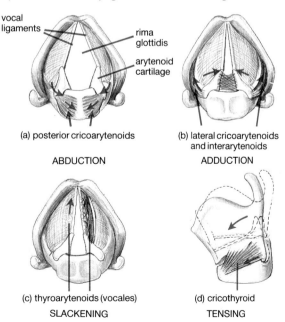

6.6.9
Movements of vocal folds.

Strap muscles of the front of the neck (6.2.1, 6.6.10)

Examine the muscles which lie anterior to the larynx and trachea and are attached to the thyroid cartilage and hyoid bone. These muscles help to control the position of the larynx and are collectively known as **'strap muscles'** from their narrow, strap-like appearance. Their lower parts lie partly under cover of sternomastoid.

Each **sternohyoid** arises from the posterior surface of the manubrium and medial end of the clavicle and extends upward to be inserted into the lower border of the body of the hyoid.

Lying in the same plane and immediately lateral to sternohyoid locate the superior belly of **omohyoid**; its inferior belly arises from the scapula close to the scapular notch and joins the superior belly at an angle through a very short intermediate tendon held to the deep surface of sternomastoid by a fibrous tissue pulley. Reflect sternohyoid to find **sternothyroid**, which arises on the posterior surface of the manubrium and inserts into the oblique line on the lamina of the thyroid cartilage. **Thyrohyoid** continues upward from the oblique line in the same plane to attach to the lower border of the greater horn of the hyoid. All the infrahyoid muscles depress the hyoid and therefore the larynx when they contract.

Locate **geniohyoid** as it passes upward from the body of the hyoid, deep to mylohyoid, to reach the inferior genial tubercles on the inner aspect of the mental symphysis. With other muscles, it pulls the hyoid bone upward and forward during swallowing.

Qu. 6G   *Which other muscles elevate the hyoid bone?*

All the strap muscles are supplied from a thin looped nerve, the **ansa cervicalis** (p. 129). Its upper root (ansa hypoglossi) branches from the hypoglossal nerve but contains fibres originating from C1; its lower root, which supplies all the strap muscles below the oblique line on the thyroid cartilage, contains fibres from the cervical plexus C2 and C3.

Protection of the airway during swallowing (see 6.5.8)

During swallowing the nasal airway is closed off from the mouth by elevation of the soft palate (p. 68); and the larynx and trachea are functionally separated though not entirely closed off from the pharynx. The constrictor muscles of the pharynx, stylo- and palatopharyngeus, and the suprahyoid strap muscles contract to raise the larynx. At the same time, the tongue is pulled backward by styloglossus. The epiglottis, which lies between the tongue and larynx, is tilted downward over the rim of the laryngeal opening by these changes and by contraction of the aryepiglottic muscles which, with the interarytenoid muscles, reduce the size of the inlet to the larynx. By these movements, swallowed food and liquids are diverted away from the laryngeal opening. In addition, the glottis is closed by adduction of the vocal folds by the lateral cricoarytenoid and interarytenoid muscles. Occasionally, these movements are not fully co-ordinated and food or liquids 'go down the wrong way'.

Movements of the larynx during variation in respiration (6.6.9, 6.6.11)

The narrowest part of the airway is the glottis. During quiet breathing the vocal folds lie in a slightly abducted position due to tonic contraction of the posterior cricoarytenoid muscles. During deep breathing the vocal folds are forcibly abducted, and the glottis held widely open by increased contraction of posterior cricoarytenoid and relaxation of the interarytenoid muscles. Remember that the posterior cricoarytenoids are the only muscles which can abduct the vocal folds; they do this both by rotating the arytenoids so that their vocal processes move laterally, and also by pulling the arytenoids apart down the sloping shoulders of the posterior arch of the cricoid cartilage. Paralysis of both posterior cricoarytenoids by interruption of their nerve supply leads to flaccid closure of the glottis and suffocation. **Coughing** is used to clear the airway and is generated by forcible exhalation against a closed glottis (**6.6.11b**) which is then suddenly opened.

Qu. 6H   *Which nerve supplies posterior cricoarytenoid?*

Movements of the larynx during phonation (6.6.9, 6.6.11c)

During phonation the larynx produces a column of air vibrating at an audible frequency. It therefore controls the pitch of the sound produced. The lips and tongue produce the consonants and vowel sounds from this basic pitch. To produce a vibrating column of air, expired air is forced past vocal folds that are adducted to produce a slit-like glottis. The vocal processes of the

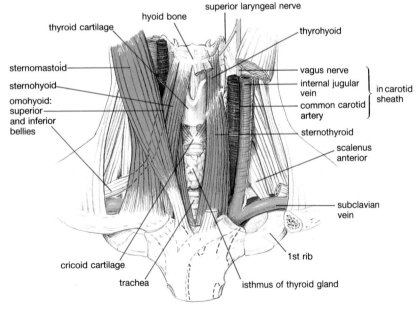

thyroid cartilage
hyoid bone
superior laryngeal nerve
sternomastoid
sternohyoid
omohyoid: superior and inferior bellies
thyrohyoid
vagus nerve
internal jugular vein
common carotid artery
} in carotid sheath
sternothyroid
scalenus anterior
subclavian vein
cricoid cartilage
trachea
isthmus of thyroid gland
1st rib

6.6.10
Infrahyoid strap muscles.

arytenoids are medially rotated and thus adducted by contraction of the lateral crico-arytenoid muscles, and the interarytenoid muscles adduct the two arytenoid cartilages posteriorly.

The **pitch** of a vocal sound is determined by the tension of the vocal folds: high-pitched sounds are produced when the vocal folds are taut; low-pitched sounds when the folds are relaxed. The vocal folds are made taut and lengthened by bilateral contraction of cricothyroid.

Qu. 6I *How is cricothyroid able to tense the vocal folds?*

Relaxation of the vocal folds is caused by contraction of the thyroarytenoid muscles which approximate the thyroid and arytenoid cartilages, thereby relaxing the vocal ligament. Some fibres of vocalis insert into the vocal fold at different points along its length and are thought to produce different vocal timbres by finely adjusting the tension at different points along the vocal folds. Whisper a few words and note that, in addition to the volume of sound being much less than in normal speech, there is a greater escape of air through the larynx. This results from adduction of the vocal folds to give a slit-like glottis (as for speech) combined with separation of arytenoid cartilages posteriorly; most of the expired air then passes between the cartilages and only a minor part passes between and vibrates the vocal folds.

Qu. 6J *Which laryngeal muscles would be contracted and which relaxed during whispering?*

Blood supply to the larynx and trachea (6.6.12)

The vessels of the larynx and cervical part of the trachea are essentially divided into an upper and lower group. The **arteries** are laryngeal branches of the superior thyroid artery (a branch of the external carotid artery) and inferior thyroid artery (a branch of the subclavian artery). Identify laryngeal branches of the superior thyroid artery which pierce the thyrohyoid membrane to gain the interior of the larynx; those from the inferior thyroid artery enter the larynx at its lower margin. **Veins** draining the larynx and trachea run with the arteries into, respectively, either the internal jugular vein or left brachiocephalic vein. The **lymphatics** run with the blood vessels to the deep cervical nodes along the internal jugular vein, and to nodes around the trachea.

Nerve supply to the larynx and trachea (6.6.13)

The sensory, motor, and secretomotor nerves to the larynx and trachea are all derived from the vagus nerve (p. 127), the motor fibres originating in the cranial accessory nerve (XI); sympathetic fibres reach it along the blood vessels.

Sensation to the larynx *above* the vocal folds is supplied by the **internal laryngeal** branch of

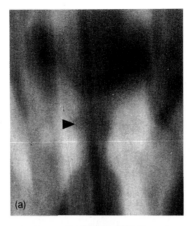
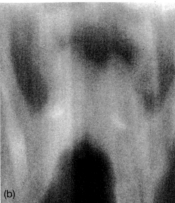
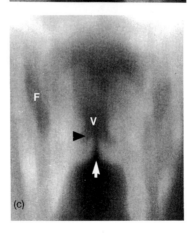

6.6.11
Frontal tomograms of larynx. (a) During quiet breathing; (b) before a cough; (c) during phonation.

the superior laryngeal nerve. Locate the internal laryngeal nerve as it passes between the middle and inferior constrictors of the pharynx and pierces the thyrohyoid membrane. Sensation *to* and *below* the vocal folds is supplied by the **recurrent laryngeal nerve** which enters the larynx by passing upward beneath the lower border of inferior constrictor.

The **motor** supply to all the intrinsic muscles of the larynx except cricothyroid is derived from the **recurrent laryngeal nerve**; cricothyroid is

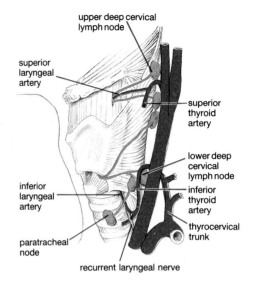

6.6.12
Arterial supply and lymphatic drainage of larynx.

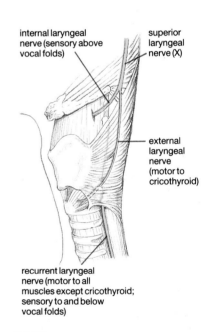

6.6.13
Nerve supply to larynx.

supplied by the **external laryngeal** branch of the superior laryngeal nerve. Identify the superior laryngeal nerve, which passes behind the carotid sheath toward the larynx and then divides into internal and external laryngeal branches.

Thyroid gland and parathyroid glands (6.6.14)

Development of the thyroid and parathyroid glands

The **thyroid gland** develops from an endodermal diverticulum which appears in the floor of the mouth immediately caudal to the tuberculum impar (p. 22) of the developing tongue (at the foramen caecum at the apex of the sulcus between the anterior and posterior components of the tongue). The diverticulum develops into a tubular duct which grows caudally anterior to the developing larynx, and bifurcates to form bilateral aggregations of cells which reorganize to form thyroid follicles within two lateral lobes. Thyroid tissue may unusually develop anywhere along the line of migration from the tongue, and the gland may also migrate too far caudally and reach the anterior mediastinum of the chest.

Qu. 6K *At what sites would enlargement of ectopic thyroid tissue create a mechanical problem?*

6.6.14
Thyroid gland and its blood supply.

The superior and inferior **parathyroid glands** derive from the endoderm of the 4th and 3rd pharyngeal pouches, respectively. Caudal migration of the thymus, which also forms from the 3rd pharyngeal pouch, carries its associated parathyroid below that derived from the 4th pouch. The parathyroids usually become embedded in the posterior surface of the developing thyroid, but unusually can migrate more widely with the thymus. Calcitonin-producing **C cells of the thyroid** are derived from the ultimobranchial body which may derive cells from the neural crest.

Prosections of the thyroid and parathyroid glands (6.1.14)

Reflect the strap muscles and examine the **thyroid gland**. Its two conical **lobes** lie on either side of the larynx and trachea, united near their base by an **isthmus** anterior to the 2nd and 3rd tracheal rings. The upper pole of each lobe extends upward to the oblique line on the lamina of the thyroid cartilage; the lower pole downward to about the 4th or 5th tracheal ring. Try to find the **parathyroid glands**: four small, yellowish, ovoid structures situated between the posterior border of the lower part of the lobes of the thyroid and its capsule. The thyroid and parathyroids are ensheathed in the **pretracheal fascia**, a thin but firm layer attached to the cricoid cartilage.

Qu. 6L *How could you easily determine whether a lump in the neck was associated with the thyroid gland?*

Note that sternothyroid lies immediately anterior to the thyroid gland; its insertion into the oblique line of the thyroid lamina prevents any enlargement of the gland from extending upward. Cricothyroid and fibres of middle constrictor which are attached to the oblique line of the thyroid lamina lie deep to the thyroid. The posterior aspect of each lateral lobe impinges on the carotid sheath and, between the gland and the sheath, branches of the inferior thyroid artery pass to the lower pole of the thyroid. Among these branches locate the recurrent laryngeal nerve running upward between the trachea and oesophagus.

Qu. 6M *If the recurrent laryngeal nerve of one side were damaged during the clamping of inferior thyroid vessels for thyroid surgery, what would be the result?*

Blood and nerve supply of the thyroid and parathyroid glands (6.6.14)

Locate the **superior thyroid artery** which is usually the first branch of the external carotid artery. It characteristically loops upward and gives superior laryngeal branches before passing down to the upper pole of the thyroid.

Now find the **inferior thyroid artery**, a branch of the thyrocervical trunk of the subclavian artery (p. 104), which runs upward and usually divides into a number of branches before entering and supplying the lower pole of the gland where it is in close relationship with the recurrent laryngeal nerve. Very often a **thyroidea ima** artery can be found which arises from the arch of the aorta and runs upward in front of the trachea to supply the isthmus. There is a good anastomosis between all the arteries supplying the gland. Veins within the thyroid form **superior, middle, and inferior thyroid veins.** The superior and middle drain into the internal jugular vein; the inferior into the left brachiocephalic vein. The lymphatic vessels draining the gland end in the thoracic duct and the right lymphatic duct.

The **nerve supply** to the thyroid gland and to its blood vessels is derived from **postganglionic fibres** of the **cervical sympathetic ganglia.**

C. Radiology

Examine the lateral radiograph of the larynx (**6.6.15**) and note that the oropharynx (O), laryngeal vestibule (V), and trachea (T) are readily identifiable. The glottis can also be identified by the small amount of air in the sinus of the larynx (arrow), which separates the vocal fold inferiorly from the vestibular fold ('false' vocal cord) superiorly. Note the epiglottis (arrow head), running upward from the base of the vestibular fold anteriorly to lie against the posterior surface of the tongue. A small amount of air is trapped in the valleculae (curved arrow) at the level of the hyoid bone. The laryngeal skeleton ossifies in a variable pattern and is frequently only identifiable as an irregular area of ossification, most frequently in the posterior part of the cricoid (C).

Look at the frontal tomograms of the larynx; in **6.6.11a** the subject is breathing quietly and the vocal fold is fully abducted. The side wall of the airway (arrow) has a smooth contour and the laryngeal sinus is shallow (arrowhead). On phonation (6.6.11c) the vocal folds are adducted and the glottis narrowed (arrow). The laryngeal sinuses (ventricle) are easily seen (arrow head). Note that air in the pyriform fossa (F) is separated from air in the laryngeal ventricle (V) by the aryepiglottic fold. **6.6.11b** was taken during a Valsalva manoeuvre, i.e. when pressure in the trachea is raised against a closed glottis (this occurs normally when coughing or increasing abdominal pressure). Note that the vocal folds are fully adducted and the laryngeal ventricle obliterated.

6.6.16 is a scintigram of the thyroid. The gland takes up technetium-99m as it does iodine-123 and the uptake of radionuclide can be monitored. The right lobe and isthmus (arrow) have a normal pattern of uptake. The left lobe shows little activity because it is occupied by a nonfunctioning benign cyst.

6.6.17 is a CT of the thyroid gland (T) at the level of T1 (note the proximal part of the 1st rib). Normal right and left lobes can be seen on either side of the trachea; they appear dense in

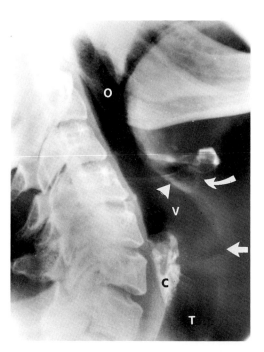

6.6.15
Lateral radiograph of larynx.

6.6.16
Scintigram of thyroid gland.

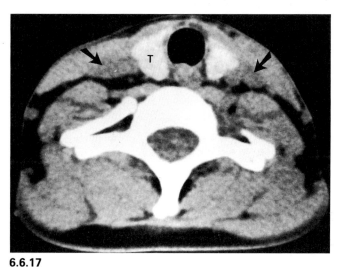

6.6.17
CT of neck showing thyroid gland.

relation to surrounding tissue because of the iodine content of the thyroid.

Qu. 6N *To what structures are the arrows in* **6.6.17** *pointing?*

Requirements:

Articulated skeleton, hyoid bone.

Prosections of larynx *in situ* in the neck; hemisected larynx; larynx opened dorsally to show interior features; cartilages and ligaments of larynx; muscles of larynx; blood and nerve supply of larynx; prosection of thyroid gland (parathyroid glands) and trachea.

Models of the larynx (if available).

Radiographs of larynx and thyroid.

Laryngoscopic photographs of vocal folds at rest and during phonation.

Seminar 7

Interior of the skull

scalp
connective tissue
muscle.
fat
periosteum

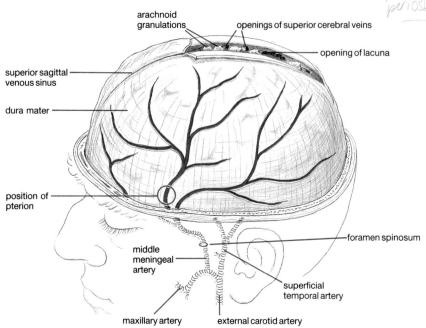

6.7.1
Dura mater and meningeal vessels.

(labels on figure 6.7.1:)
arachnoid granulations
openings of superior cerebral veins
opening of lacuna
superior sagittal venous sinus
dura mater
position of pterion
foramen spinosum
middle meningeal artery
superficial temporal artery
maxillary artery
external carotid artery

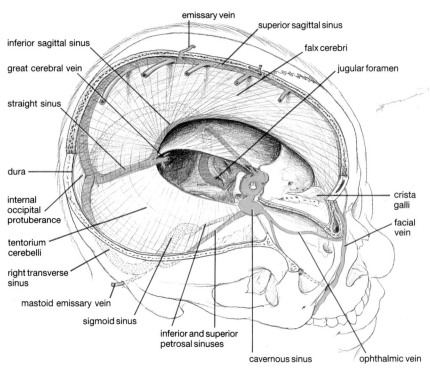

6.7.2
Dural venous sinuses.

(labels on figure 6.7.2:)
emissary vein
superior sagittal sinus
inferior sagittal sinus
falx cerebri
great cerebral vein
jugular foramen
straight sinus
dura
internal occipital protuberance
crista galli
tentorium cerebelli
facial vein
right transverse sinus
mastoid emissary vein
sigmoid sinus
inferior and superior petrosal sinuses
cavernous sinus
ophthalmic vein

The **aim** of this seminar is to study the interior of the skull, its meningeal linings, and the venous sinuses within the dura; and the vessels and nerves that enter and leave the skull. Before doing so, review the relevant parts of the skull (p. 29).

A. Prosections of the interior of the skull

The meninges (6.7.1–6.7.4)

Three membranes, the **meninges**, surround the brain within the skull: an outer, tough, fibrous **dura mater** which is attached to the skull, a more delicate fibrous **arachnoid mater** which lines the inner surface of the dura, and the thin **pia mater** which is attached to the brain tissue. **Cerebrospinal fluid** fills the space between the arachnoid and pia mater to support and protect the central nervous system.

Examine a head from which the skull cap has been removed (**6.7.1**) to expose the **dura mater**. Its two layers, which form the inner periosteum of the skull and a meningeal covering for the brain (**6.7.2**), are fused except where they diverge to enclose venous sinuses. Between the dura and the skull is a potential space, the **extradural space**, where blood collects if meningeal vessels are ruptured, but tight adherence of the dura to skull sutures limits the extension of any such collection of fluid. Try to find some branches of the **middle meningeal vessels** (**6.7.1**) which ramify in the extradural space.

Examine a prosection in which at least one-half of the cerebrum has been removed. Identify the **falx cerebri** (**6.7.2**), a sickle-shaped extension of the inner layer of the dura which lies between the two hemispheres. Anteriorly, it is attached to the crista galli; its upper margin is attached to the skull cap in the midline as far back as the internal occipital protuberance; its lower margin is free anteriorly but attached posteriorly to another extension of the inner layer of dura, the tentorium cerebelli. The **tentorium** (**6.7.2**) separates the occipital lobes of the cerebral hemispheres from the cerebellar hemispheres, which lie in the posterior cranial fossa. Trace the attachments of the tentorium: from its midline attachment to the falx cerebri and internal occipital protuberance, it passes horizontally around the inner aspect of the occipital and squamous temporal bones to the upper margin of the petrous temporal bone, and then forward along this to the posterior and anterior clinoid processes of the sphenoid. It slopes upward like the roof of a tent toward its attachment to the falx, and has a crescentic free anterior border which lies around the brainstem. Between the anterior and posterior clinoid processes, the dura roofs over the pituitary fossa.

The **arachnoid mater** (**6.7.4**) is a delicate fibrous membrane which closely lines the dura mater, thus the 'subdural space' (between dura and arachnoid) is only a potential space. Within the arachnoid lies the **subarachnoid space** which is finely trabeculated like a spider's web (**arachne = spider**) and filled with **cerebrospinal fluid** (**6.7.3**). All vessels and nerves passing to or from the brain must cross the subarachnoid space, and many have an extensive intracranial course.

Qu. 7A *If an aneurysm on an intracranial artery burst, in which space would the blood collect, and how might this be detected?*

The **pia mater** (**6.7.4**) is a delicate vascular meninge closely adherent to the brain tissue. It encloses the brain, spinal cord, and nerve roots and is prolonged around small arteries as they enter the brain tissue.

Cerebrospinal fluid (CSF) is secreted into a series of linked cavities (**ventricles**) within the brain by **choroid plexuses** which are composed of capillaries surrounded by a secretory epithelium. The CSF circulates through the ventricles and leaves the ventricular system by passing through openings in the roof of the fourth ventricle to reach the subarachnoid space of the posterior cranial fossa. It circulates around and supports the brain and spinal cord and then diffuses back into the venous system, in large part via **arachnoid villi** and **arachnoid granulations** (**6.7.4**) which project through the dura into the cranial venous sinuses.

Cranial venous sinuses (6.7.2, 6.7.5)

Examine a specimen in which the endothelium-lined **venous sinuses** between the outer and inner layers of the cranial dura have been exposed. In the median plane just beneath the sagittal suture of the skull locate the large **superior sagittal sinus** which extends from the crista galli anteriorly to the internal occipital protuberance where it joins (usually) the right transverse sinus. Explore the sinus and locate its lateral recesses, the openings of superior cerebral veins and the arachnoid granulations, which open into it and create rounded indentations (foveolae) on the skull cap lateral to the sinus (**6.7.4**).

Qu. 7B *On what factors does a passive diffusion of CSF into venous blood depend?*

Qu. 7C *The anterior fontanelle closes at about 18 months of age. What deductions about body fluid volume might be made by palpating the fontanelle?*

The brain can move a little within the cerebrospinal fluid, so that injuries which cause severe

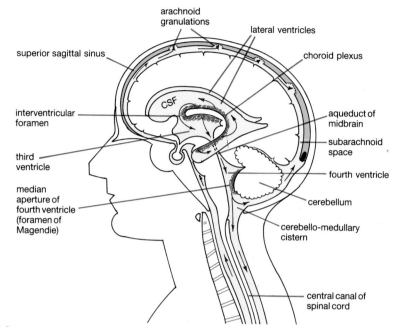

6.7.3
Circulation of cerebrospinal fluid.

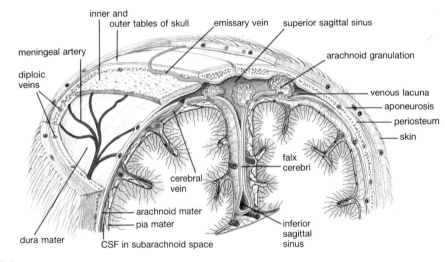

6.7.4
Coronal section through cranium to show meninges and arachnoid granulations.

acceleration or deceleration of the head may cause traction on thin-walled superior cerebral veins as they enter the superior sagittal sinus (**6.7.5**). This can lead to rupture of the veins and extensive bleeding into the subdural space.

Trace the superior sagittal sinus into the right **transverse sinus** which is situated between the two layers of dura within the attached margin of the tentorium cerebelli. At the posterior aspect of the petrous temporal bone it swings downward and medially to form the S-shaped **sigmoid sinus** which grooves the bone, and exits through the jugular foramen (**6.7.6**).

In the inferior margin of the falx cerebri locate the **inferior sagittal sinus**. This is joined posteriorly by veins draining both the interior and exterior of the brain to form the **straight**

sinus which lies at the junction between the falx and the tentorium and passes backward to the interior occipital protuberance where it turns, usually to the left, to form the left **transverse sinus**.

Identify the **cavernous sinuses** which are situated bilaterally on either side of the body of the sphenoid bone. The cavernous sinuses communicate with and drain veins of the orbit and deep parts of the face and also the lateral

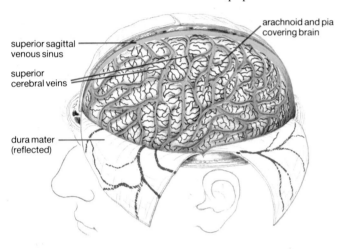

6.7.5
Entry of superior cerebral veins into superior sagittal sinus.

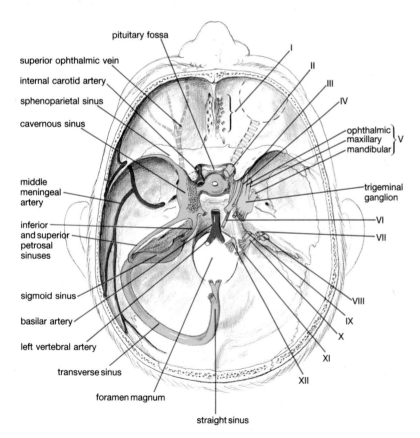

6.7.6
Interior of skull base; vessels and nerves.

aspects of the cerebral hemispheres and the pituitary gland. The two cavernous sinuses are interconnected around and beneath the pituitary, and drained posteriorly by two smaller petrosal sinuses. Each **superior petrosal sinus** lies within the margin of the tentorium attached to the upper margin of the petrous temporal bone, and connects the cavernous sinus to the transverse sinus; each **inferior petrosal sinus** passes downward behind the medial extremity of the petrous temporal bone to enter the anterior aspect of the jugular foramen and drain into the sigmoid sinus with which it forms the **internal jugular vein**.

Structures entering and leaving the cranial cavity (6.7.6)

The cranial cavity is largely occupied by the brain. The supratentorial compartment, which includes the anterior and middle cranial fossae, contains the two cerebral hemispheres. The anterior cranial fossae contain the frontal lobes; the middle cranial fossae contain the temporal lobes; the parietal lobes lie superiorly beneath the vault of the skull; and the occipital lobes lie posteriorly above the tentorium. Beneath the tentorium, the cerebellar hemispheres occupy the posterior cranial fossa. Centrally lies the brainstem, which connects the thalamus and hypothalamus of the cerebrum with the spinal cord. The brainstem comprises the midbrain, which lies at the level of the free edge of the tentorium (tentorial notch), and below that, the pons and medulla oblongata of the hindbrain.

Anterior cranial fossa
In the anterior cranial fossa locate the **olfactory bulbs** which lie over the cribriform plates. They receive olfactory nerve (I) fibres from the nose and their **olfactory tracts** extend backward on either side to join the temporal and frontal lobes.

Middle cranial fossa
At the posterior margin of the anterior cranial fossa, the **optic nerves** (II) enter the cranial cavity through the optic foramina and join to form the **optic chiasm**. Immediately behind the optic nerves locate the **internal carotid arteries** as they emerge from the upper aspect of the cavernous sinus. On an isolated skull trace the course of the artery: it passes through the carotid canal in the petrous temporal bone, enters the cranial cavity through the upper part of the foramen lacerum, and then curves anteriorly taking an S-shaped course on the side of the body of the sphenoid (i.e. the medial wall of the cavernous sinus). As it emerges from the cavernous sinus at the anterior clinoid process, it turns sharply backward and gives off the **ophthalmic artery (6.7.6, 6.7.7)** which accompanies the optic nerve through the optic foramen to the orbit.

Qu. 7D *If the walls of the internal carotid artery formed an aneurysm which ruptured into the cavernous sinus, what might result?*

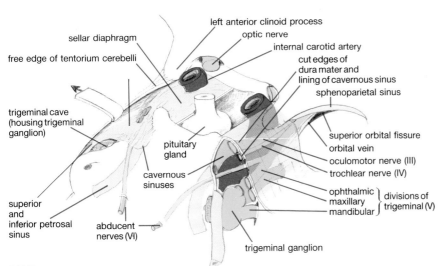

6.7.7
Cavernous sinus and related structures.

Pituitary gland (for development of the pituitary see p. 19)

Immediately posterior to the optic chiasm a small sheet of dura mater (sellar diaphragm), attached to the anterior and posterior clinoid processes of both sides, roofs the pituitary fossa and is pierced by the stalk (**infundibulum**) of the pituitary gland (**6.7.7**). The **pituitary gland** lies within its fossa (sella turcica) on the sphenoid bone. It consists of two main parts, the **neurohypophysis** (neural lobe, posterior pituitary) which is a downward extension of the hypothalamus containing terminals of hypothalamic neuroendocrine neurons and the **adenohypophysis** (anterior lobe) which consists of endocrine cells of which some extend up around the infundibulum to form the tuberal part of the gland. A narrow cleft and a few endocrine cells (intermediate lobe) separate the main anterior and posterior parts of the gland.

　Connecting the hypothalamus to the anterior pituitary is a system of portal veins (**hypothalamo–hypophysial portal vessels**) which transmit to the anterior pituitary regulatory hormones produced by the hypothalamus. These important vessels are, unfortunately, too small to find in a formalin-fixed cadaver. On either side of the pituitary and sella turcica lie the internal carotid arteries and the cavernous sinuses. The internal carotid arteries send small (inferior hypophysial) branches to the posterior lobe, but the base of the hypothalamus and pituitary stalk, from which the portal veins originate, receives its arterial supply from (superior hypophysial) branches of the middle cerebral branch of the internal carotid artery (p. 106). The cavernous sinuses drain the pituitary gland and its endocrine secretions into the systemic bloodstream.

Qu. 7E *What are the functional connections between the pituitary gland and the hypothalamus?*

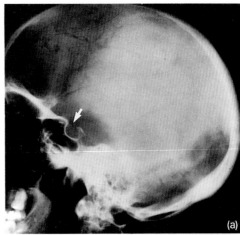

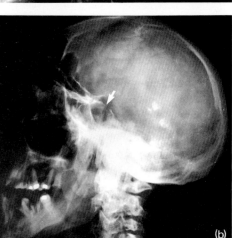

6.7.8
Radiographs of pituitary fossa (arrow): (a) normal; (b) enlarged by pituitary tumour.

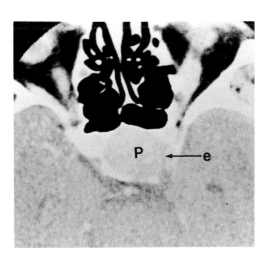

6.7.9
CT showing enlarged pituitary gland (P) eroding clinoid processes (e).

As tumours of the pituitary gland enlarge, they can erode the clinoid processes of the sphenoid bone and enlarge the pituitary fossa. Compare the pituitary fossa on radiograph **6.7.8a** (normal) with that on **6.7.8b** from a patient with a growth hormone-producing tumour that has expanded the fossa and caused acromegaly which is responsible for the enlarged lower jaw. Examine also CT **6.7.9** and identify the erosion of the clinoid processes seen in this horizontal section. Pituitary tumours may also enlarge upward through the dural roof of the pituitary fossa.

Qu. 7F *What structures could an enlargement of the pituitary press on, and what might be the consequences?*

Within the middle cranial fossa, the three cranial nerves which supply the extrinsic muscles of each eye pass through the walls of the cavernous sinus *en route* to the orbit. Locate the **oculomotor nerve** (III) as it emerges from the ventral aspect of the brainstem between the two cerebral peduncles to pass forward in the subarachnoid space. At the posterior wall of the cavernous sinus it pierces the dura and runs forward in the lateral wall of the sinus. The slender **trochlear nerve** (IV) emerges from the dorsum of the midbrain and winds laterally around the brainstem to reach the cavernous sinus where it pierces the dura immediately behind the oculomotor nerve. Within the lateral wall of the sinus the two nerves cross as they pass into the orbit through the superior orbital fissure. Trace the **abducent nerve** (VI) as it leaves the ventral aspect of the brainstem between the pons and medulla and passes upward to pierce the dura covering the basisphenoid. As it crosses the apex of the petrous temporal bone, it enters the medial wall of the cavernous sinus where it lies immediately lateral to the internal carotid artery before it, too, passes forward into the orbit through the superior orbital fissure.

Qu. 7G *Raised intracranial pressure can cause traction on the abducent nerve(s) and paralysis of the muscle it supplies. What would be the result (see p. 117)?*

Locate the large **trigeminal (V) nerve** (**6.7.6, 6.7.7**) which emerges from the lateral aspect of the brainstem (pons) in the posterior cranial fossa to enter a CSF-filled recess in the dura (trigeminal cave) on the upper surface of the apex of the petrous temporal bone. Here the nerve expands to form the flattened **trigeminal ganglion**.

Qu. 7H *The trigeminal ganglion is the equivalent of a dorsal root ganglion. Of what, therefore, is it comprised?*

The three major sensory divisions of the trigeminal nerve enter the anterior aspect of the ganglion but the motor root passes deep to it. The **ophthalmic division** of the nerve passes forward within the lateral wall of the cavernous sinus and divides into branches (p. 117) which enter the orbit through the superior orbital fissure. The **maxillary division** runs forward on the floor of the middle cranial fossa at the lateral extremity of the cavernous sinus and passes through the foramen rotundum and on to the floor of the orbit. The **mandibular division** leaves the skull with the motor root through the foramen ovale (**6.7.6**) to reach the infratemporal fossa.

Immediately lateral to the foramen ovale identify the **middle meningeal artery** as it enters the skull through the foramen spinosum with the recurrent sensory branch of the mandibular nerve (V) which supplies the dura of the middle cranial fossa. The middle meningeal vessels, which supply the dura and cranial bones, lie in the extradural space of the anterior and middle cranial fossa, and groove the skull bones. Other small meningeal vessels and branches of the trigeminal and vagus nerves serve the anterior and posterior fossae. On an isolated skull, trace the course of the middle meningeal artery and its anterior and posterior branches. Note that the anterior branch runs beneath the pterion (p. 107) often in a bony canal where the bone can be quite thin.

Qu. 7I *If the middle meningeal artery were ruptured by a blow in the region of the pterion, where would blood collect?*

Posterior cranial fossa
Identify the **facial nerve** (VII) and **vestibulocochlear nerve** (VIII) as they leave the lateral aspect of the brainstem and pass across the cerebello-pontine angle with the labyrinthine artery to enter the internal auditory meatus on the posterior aspect of the petrous temporal bone. Locate the **glossopharyngeal nerve** (IX), **vagus nerve** (X), and **cranial root of the accessory nerve** (XI) as they pass from the lateral aspect of the medulla to leave the skull through the jugular foramen. Fibres of the **spinal root of the accessory nerve** (XI) arise as a series of rootlets from the upper cervical (C2, 3, 4) part of the spinal cord, and pass upward through the foramen magnum. They briefly join the cranial root of the accessory nerve within the jugular foramen, but then diverge from it again as they pass to supply sternomastoid and trapezius in the neck. The cranial root of the accessory joins the vagus nerve. Identify rootlets of the **hypoglossal nerve** (XII) which emerge from the anterior aspect of the brainstem and leave the skull through the hypoglossal canal (anterior condylar canal) in the occipital bone. Finally, identify the structures passing through the foramen magnum. These are the **medulla of the brainstem** which becomes continuous with the spinal cord, the two **vertebral arteries** which pierce the posterior atlanto-occipital membrane and pass through the foramen magnum to unite on the anterior aspect of the brainstem as the basilar artery, and the **spinal roots of the accessory nerve** (XI).

B. Radiology

Examine an external carotid angiogram (see **6.10.12**) and locate the middle meningeal artery and its branches.

On A–P and lateral internal carotid angiograms (see **6.10.11a,b**) trace the course of the internal carotid artery as it passes through the petrous temporal bone and cavernous sinus (the carotid 'syphon'), noting its relation to the pituitary fossa. The venous phase angiogram (see **6.10.11c**) was taken when radio-opaque material introduced into the internal carotid artery had reached the veins. Identify the straight sinus and the transverse and sigmoid sinuses.

Examine the CT **6.7.10** through the base of the skull; the left side is intact and should be compared with the right side which has been fractured.

Qu. 7K *Which nerves are most likely to have been damaged by this injury?*

Requirements:

Skull and skull cap.

Prosections of the cranium with skull cap removed to show dura; with dura removed to show cerebral hemispheres; with the cerebral hemisphere of one side removed to show the falx cerebri; with cerebral and cerebellar hemispheres removed and dura incised to show venous sinuses; prosections of floor of the cranial cavity showing cranial nerves, and the cavernous sinus and its contents.

Radiographs of skull; internal and external carotid arteriograms and internal carotid arteriogram in venous phase to show dural sinuses.

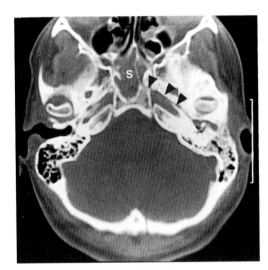

6.7.10
CT showing fracture (arrowheads) of base of skull and opacity of sphenoid air sinus (S) due to collection of blood. Scale bar, 5 cm.

Seminar 8

Eye and orbit

The **aim** of this seminar is to study the eye, including its gross structure and movements, its nerve and blood supply, and its surroundings within the orbit.

Development of the eye

(6.8.1)

In the early embryo bilateral optic sulci appear in the walls of the forebrain while the rostral neuropore is still widely open. After closure of the neuropore, each sulcus bulges and extends laterally to form a diverticulum, the lumen of which is continuous with that of the developing forebrain. The proximal end of the diverticulum elongates and constricts to form the **optic stalk**, while its peripheral end expands to form an optic vesicle. The distal part of the vesicle invaginates into its proximal part thus occluding the cavity and forming a double-layered **optic cup**. The cavity is not completely obliterated and is the site of pathological retinal detachment in the adult. The outer layer of the cup becomes the **pigmented layer of the retina**; the apposed inner layer forms the neural elements of the **retina**. Within the neural layer of the retina, the photoreceptors and other neurons differentiate, and the axons (neurites) of the innermost layer of neurons grow backward through the optic stalk to form the **optic nerve**. More anteriorly, the pupillary part of the neuroectodermal optic cup forms the muscles of the iris, and the epithelium of the iris and ciliary bodies (but not the ciliary muscle which is derived from mesoderm).

At an early stage in this process the presence of the optic vesicle induces in the overlying surface ectoderm a thickening, the **lens placode**, which invaginates to form a **lens vesicle**. This separates from the covering ectoderm to form the **lens** primordium which comes to lie within and practically fills the optic cup. On the under-surface of the optic cup and stalk, a **choroidal fissure** forms into which **choroidal vessels**, including the **hyaloid artery** which supplies the lens, grow to supply the inner structures of the eyeball. As the lips of the fissure grow together they fuse, thus isolating the vessels within the interior of the eyeball. Rarely, the choroidal fissure fails to close, leaving an inferior defect in the iris and choroid (**coloboma**). Vascularization of the lens capsule normally regresses before birth; failure to do so is associated with certain forms of congenital cataracts. The space between the developing lens and retina increases in size and becomes filled with jelly-like **vitreous humour**, while the mesenchyme between the lens and surface ectoderm develops a cleft and becomes filled with **aqueous humour**. Around the optic cup and lens vesicle, the mesenchyme forms two connective tissue layers, the outer **sclera** and inner, vascular **choroid**, which are in continuity with the meninges; the anterior part of the choroid forms the ciliary body, ciliary processes, and ciliary muscle.

The eyelids develop from small cutaneous folds which extend over the surface of the eye. These grow together and unite, and only separate again toward the end of the intrauterine life.

The **lacrimal gland** arises as a series of ectodermal buds which grow into the mesenchyme between the upper eyelid and eyeball,

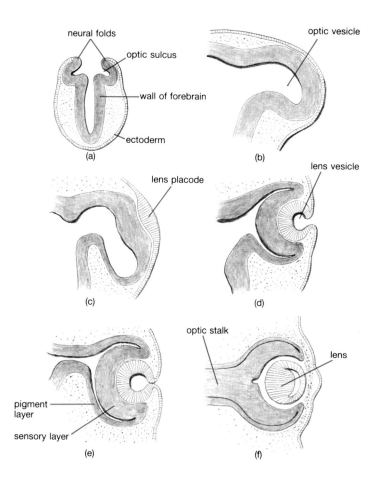

6.8.1
Development of eye.

laterally. The **lacrimal sac** and **nasolacrimal duct** arise from canalization of a cord of ectodermal cells buried at the junction of the lateral nasal and maxillary process (p. 105).

A. Living anatomy

Examine your own or your partner's eyes; their appearance depends on both the eyeball and eyelids (**6.8.2a**). Between the upper and lower lids (palpebral fissure) can be seen part of the anterior aspect of the eyeball, including the **sclera** (the 'white' of the eye) which is continuous with the transparent **cornea** and, within the eye, the pigmented **iris** which surrounds the **pupil**.

It is important to be able to recognize a normal appearance of the eye and to appreciate the range of ethnic differences (**6.8.2b**) which include the 'set' of the eyes (i.e. whether the eyelids are horizontally positioned or oval and slanted) and the occurrence of an epicanthic fold (seen in orientals but also in caucasians with Down's syndrome).

Note that the upper lid normally just overlaps the iris, whereas sclera is commonly seen between the lower lid and iris. Sclera visible between the iris and the upper lid may indicate either retraction of the lid or protrusion of the globe.

Qu. 8A *Comment on the appearance of the eyes in* **6.8.3**.

Examine the upper and lower eyelids (palpebrae), two thin, flexible folds which, with the lashes, protect the front of the eye. Ask your partner to look up and then down and note how the eyelids move with the eyeball. Very thin skin covers the outer surface of the eyelids, and eyelashes arise from their free margin. Two **tarsal plates** form a connective tissue 'skeleton' for the lids and their inner surface is lined by **conjunctiva**.

If a foreign body becomes trapped between the eyeball and the lids, it causes great pain, and must be removed. Get your partner to look down and gently but firmly take the upper lid and evert it so that the tarsal plate is exposed (ask a clinically-experienced staff member to show you how to do this). Between the tarsal plate and conjunctiva, look for the vertically oriented **tarsal glands**, modified sebaceous glands the ducts of which open on to the margin of the eyelids. If tarsal glands become infected they cause swellings (cysts), which may require surgical excision, on the inner aspect of the (usually lower) lid. The eyelash follicles can also become infected to form 'styes' which make the outer surface of the eyelid red and swollen. Look also for the **lacrimal punctum** (**6.8.4**; see below) on the **lacrimal papilla** at the medial extremity of the lid.

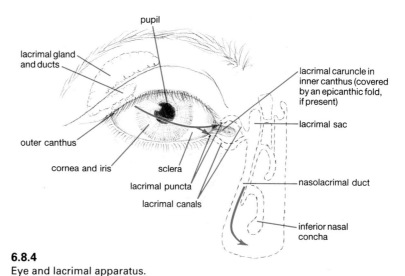

6.8.3
Exophthalmos in hyperthyroid patient.

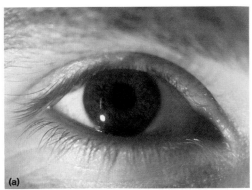

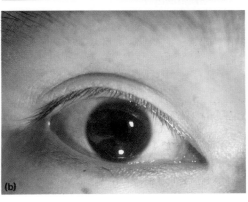

6.8.2
(a) Caucasian; (b) Mongoloid eyes.

6.8.4
Eye and lacrimal apparatus.

Now gently pull down the lower lid and identify its (smaller) tarsal plate, tarsal glands, and the lower lacrimal papilla and punctum. Examine the **conjunctiva** which lines the lid and is reflected on to the sclera and over the cornea, to form the **conjunctival sac**. The conjunctiva should be moist, and that which lines the lids is very vascular. Lacrimal secretions (tear fluid) which keep it moist are secreted by the **lacrimal gland** which lies at the upper lateral aspect of the orbit. Blinking carries the aqueous tear fluid over the eyeball and rapid evaporation of the tear fluid and drying of the exposed surface of the eyeball is prevented by the oily secretions of the tarsal glands which form a film over the tear fluid. Try to stare fixedly forward without blinking — you will find it progressively harder to prevent blinking, which occurs reflexly in response to drying. Gently touch the surface of the cornea with a wisp of cotton wool; protective reflex blinking (corneal reflex p. 141) will again be elicited.

Qu. 8B *What might result if the corneal reflex were absent?*

Lacrimal fluid is drained from the conjunctival sac by capillarity through the lacrimal puncta which normally lie in contact with the scleral conjunctiva. If the production of lacrimal fluid becomes too great, if the lids become everted, or if the drainage system becomes blocked, tear fluid will spill on to the cheeks.

Note the differences between the medial and lateral angles of the eye: at the lateral angle the two lids meet directly; at the medial angle they are separated by a small triangular space in which lies the **lacrimal caruncle**. Ask your partner to look as far laterally as possible — a semicircular fold of conjuctiva will be pulled into view by the lateral rotation of the eyeball.

Eye movements

The eyes undergo a number of different types of movement. 'Saccadic' movements rapidly direct the fovea to a target; smooth 'pursuit' movements keep the fovea on the target after it has been located; vestibulo-ocular reflexes and optokinetic movements allow the gaze to be fixed on a target during movement of the head (vestibulo-ocular reflexes use afferent signals from the vestibule; optokinetic movements use signals from the retina and are involved in longer-term fixation: when the extent of head movement means that the target can no longer be fixated, then a rapid saccadic movement moves the fovea to a new target); 'vergence' movements (convergence and divergence) direct the two foveae on to targets moving towards or away from the head, and permit stereoscopic vision. All these movements are generated by central nervous control mechanisms.

Test the **range of movements** of the eyes by asking your partner to watch a moving object whilst keeping his head still. The eyeball can be moved in any direction by a combination of upward, downward, medial, and lateral movements, but never normally rotates on the visual axis. Movement is not possible beyond the point at which the pupil would be obscured by the lid. Note that, at rest, the two eyes are normally both directed towards the target and that they move together (conjugate deviation) as they follow an object which moves across the field of vision. Now ask your partner to follow an object as it is moved towards him in the midline, and watch the two eyes converge.

Examine the sclera with its covering of conjunctiva. It should appear white, but may become reddened (by infection or 'morning after the night before') or yellow (as in jaundice).

Examine the **cornea**. It should be transparent, though in older people an opaque ring (arcus senilis) may appear at the margin. It should also bulge forward with greater curvature than the remainder of the globe. The pigmented **iris** surrounds the **pupil**, and both should be symmetrical.

Shine a light into the pupil of one eye (or cover one of your partner's eyes with your hand for a minute then remove your hand) and watch the pupil constrict as light entering the eye increases.

Qu. 8C *What happens to the other pupil? The reason will become clear when you have studied the oculomotor (III) nerve (p. 116, 139).*

Ask your partner to focus on an object held at arm's length. As the object is brought closer to the head, note that, in addition to convergence of the eyes, the pupils constrict slightly for near vision (accommodation reflex; p. 140).

Ask your partner to fix his gaze on some distant point and look through the pupillary aperture to the **retina** with the help of an ophthalmoscope (ask a demonstrator to show you how to do this). Locate the **optic disc** and note its colour and reasonably distinct edges (**6.8.5a**). Note the vessels radiating from the disc and the difference between arteries and veins. Try to find the **macula** (fovea) a little lateral to the optic disc; it looks a little yellow in contrast to the red appearance of the remainder of the retina. (Note that the retina is virtually the only place in the body where blood vessels can be visualized directly.)

Finally, ask your partner to close his eyes and remind yourself of the difference between gentle closure, and 'screwing up' the eyes (p. 46). Note the thickness and extent of the eyebrows (patients with hypothyroidism tend to lose the hairs over the lateral aspect of their eyebrows).

B. Dissection and gross anatomy of the eyeball

(6.8.6, 6.8.7)

Eyeballs from human cadavers are normally not well preserved, and the eyes may even have been donated for corneal transplantation. Therefore, before studying this section, obtain an ox eye that you can dissect. Its structure is

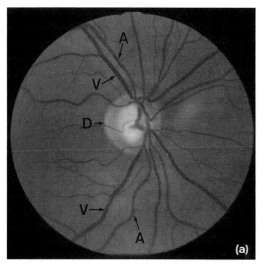

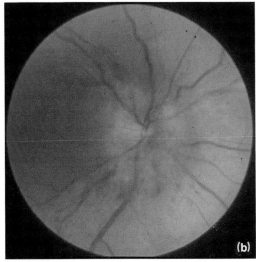

6.8.5
(a) Normal optic disc; (b) swollen/degenerating optic disc. Optic disc (D); retinal arteries (A); retinal veins (V).

essentially the same as that of the human eye, though it is much larger, and the pupil is of a different shape.

Study first the external features. The human eyeball is a sphere of approximately 2.5 cm in diameter, except that the transparent cornea bulges slightly forward. Incoming images should be focused on the macula, so that the anteroposterior dimension of the eye should be matched to the refractive power of the different media.

Qu. 8D *How will visual performance be affected if the eyeball is longer or shorter than normal relative to the power of the lens?*

Posteriorly, the **optic nerve** joins the sphere just medial to its anteroposterior axis. Within the optic nerve is the **central artery of the retina** and accompanying **veins**. Around the optic nerve is a fibrous sheath that is continuous with the dura mater; this is lined by arachnoid and separated from the nerve by an extension of the subarachnoid space. Therefore, any increase in CSF pressure (often caused by a 'space occupying lesion' within the cranial cavity) is transmitted to this sleeve of subarachnoid space and results in compression of the veins within the optic nerve. This in turn causes the retinal veins to become distended and venous congestion occurs. The resulting oedema causes the optic disc to lose its normal cupped shape and its margins become blurred. All these changes can be detected with an ophthalmoscope (**6.8.5b**).

Qu. 8E *What will happen if increased pressure on the optic nerve continues? and what would you expect the optic disc to look like?*

Forming a ring around the exit of the optic nerve are the points of entry of **ciliary nerves** and **arteries** which supply all parts of the eyeball apart from the retina, while the **veins**

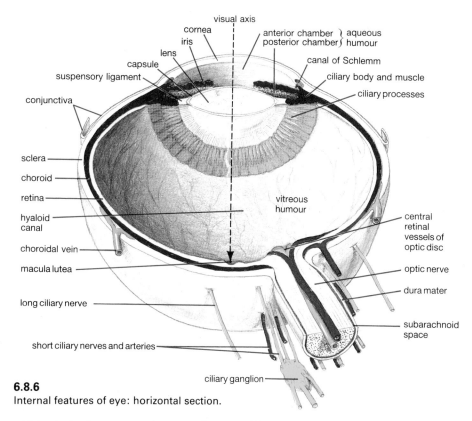

6.8.6
Internal features of eye: horizontal section.

which drain it emerge close to the coronal 'equator' of the sphere.

Chambers of the eye

Section an eyeball in the sagittal plane and identify the features in **6.8.6** and **6.8.7**. The chambers in front of and behind the iris are known respectively as the **anterior** and **posterior chambers** of the eye; they are both filled with **aqueous humour** which is secreted by **ciliary processes** of the choroid in the posterior chamber of the eye. The cavity of the eyeball behind the lens is filled with jelly-like **vitreous humour** which is enclosed in a delicate membrane, the **hyaloid membrane**. Running through the vitreous humour from the optic nerve to the back of the lens is the **hyaloid canal** which, in the fetus, transmits a branch of the central artery of the retina to the developing lens.

Coats of the eye

The eyeball has three coats. To visualize them, cut a 'window' into the outer surface of the eyeball and peel away each layer in turn.

The **sclera** or outermost fibrous protective coat forms the 'white of the eye'. It is continuous posteriorly with the dura mater of the optic nerve; anteriorly it forms the transparent **cornea** which also consists essentially of alternating lamellae of fibrous tissue. The cornea is covered anteriorly with a layer of (non-keratinized) stratified squamous epithelium which is continuous with the conjunctiva, and lined posteriorly by a layer of polygonal squamous cells which forms the endothelium of the anterior chamber. The cornea is transparent because the refractive index of all its components is the same, always provided that the anterior surface of the eye is kept moist. Most refraction of light occurs at the air-corneal interface.

Qu. 8F *If contact lenses are worn, where does most refraction of light occur?*

Examine the outer aspect of the corneo-scleral junction and note the slight change of curvature that occurs here. Within the fibrous tissue of the corneo-scleral junction lies the circular **scleral sinus (canal of Schlemm)**, into which aqueous humour of the anterior chamber drains through the trabecular lattice of fibres that comprises the wall of the sinus abutting the anterior chamber. From the scleral sinus, aqueous humour drains posteriorly into anterior ciliary veins in the choroid coat.

Examine the **irido-corneal angle** ('filtration' angle) between the anterior surface of the iris and the posterior extremity of the cornea.

Qu. 8G *If the irido-corneal angle were sufficiently narrow ('closed'), or the sinus blocked, so as to obstruct the drainage of aqueous humour, what would result? What change in the iris might ease the obstruction, and how might this be achieved?*

Just posterior to the scleral sinus, look for the tiny **scleral spur** which projects forward and inward and to which the ciliary muscle is attached.

Examine the middle coat of the eye, the **choroid**, which is vascular and pigmented. It forms a thin layer which lines the posterior 80 per cent of the eyeball and which comprises a fine capillary network and the vessels that supply it. Anteriorly, at the level of the scleral spur, it forms the **ciliary body** from which project 70–80 **ciliary processes** and to which the **suspensory ligaments** of the lens are attached (**6.8.7**). The ciliary processes are covered with an epithelium which secretes the aqueous humour.

Further anteriorly the choroid coat forms the **iris**, a pigmented muscular and vascular diaphragm which projects medially and somewhat forward in front of the lens and surrounds the central **pupil**.

The biconvex **lens** is composed of layers of ribbon-like lens fibres which together form a series of concentric layers, each of which lies at right angles to the next. The lens is nearly spherical in the fetus, but grows throughout life and, in adults, the convexity of the posterior surface is greater than that of the anterior.

Examine the way in which the lens is suspended behind the iris by its circular **suspensory ligament** which is composed of numerous fine fibres attached to the ciliary body. The tension on the suspensory ligament, and thus the curvature of the lens and its refractive power, are controlled by the **ciliary muscle**.

Qu. 8H *In old age the lens becomes much stiffer and both its surfaces become more flattened. What effect does this have on vision?*

In elderly people the lens may also become opaque to form a 'cataract'. If the opacity becomes severe the lens can be removed surgically to restore vision. Its refractive power can be replaced by spectacles, or by implantation of an artificial plastic lens. Cataracts are often yellow in colour so that, after their removal, the patient is surprised by the restoration of environmental colours, especially within the blue range.

The **retina**, which forms the inner coat of the eye, is a delicate membrane consisting of two layers. The outer layer which abuts the choroid is a single-layered epithelium of pigmented cells; they continue forward on to the ciliary processes. The pigment cells prevent lateral light scatter and back-reflection among the photoreceptors (see below).

On the inner aspect of the pigmented layer is the neural layer. The receptor cells, cones and rods, lie next to, and their photoreceptive processes are partly enclosed by, the pigment cells. The rods and cones are connected in a complex

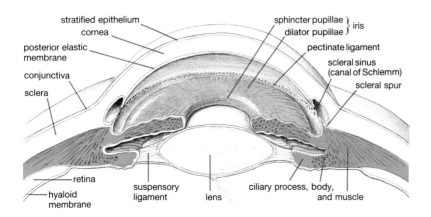

stratified epithelium
cornea
posterior elastic membrane
conjunctiva
sclera

sphincter pupillae ⎫ iris
dilator pupillae ⎭
pectinate ligament
scleral sinus (canal of Schlemm)
scleral spur

retina
hyaloid membrane
suspensory ligament
lens
ciliary process, body, and muscle

6.8.7
Coronal section of front of eye.

manner (via bipolar, horizontal, and amacrine cells) to the ganglion cells which lie next to the vitreous humour. The ganglion cells are the output neurons of the retina; their long axons sweep towards the **optic disc** where they are collected together to form the **optic nerve** (II). About 3 mm lateral to the optic disc, in the anteroposterior axis of the globe, you will find the **macula lutea** which appears yellow when viewed through an ophthalmoscope. It is the site of most acute vision; and cones are the only photoreceptors present. It is slightly hollow because the other nervous elements are displaced towards its periphery to allow light a more direct access to the receptors.

On the vitreous surface of the retina lie the **retinal arteries** and **retinal veins** and their branches. Look again at the region of the macula with an ophthalmoscope; the vessels converge toward, but do not lie over the macula.

Intrinsic muscles of the eyeball (6.8.7)

Control of pupillary diameter

The pupil contracts and dilates (from 1 to 8 mm diameter) to regulate the amount of light passing to the retina. The pupillary aperture is controlled by the balance between contraction of smooth-muscle fibres oriented in a radial — **dilator pupillae** — and circular — **sphincter pupillae** — manner. Both these muscles lie within the substance of the iris, the sphincter forming a ring just around the pupil. Try to identify their fibres by use of fine forceps. Dilator pupillae is activated by noradrenergic sympathetic nerves, whereas sphincter pupillae is supplied by cholinergic parasympathetic fibres from the ciliary ganglion (p. 116) which run with the oculomotor nerve (III).

Control of accommodation

Accommodation is controlled by the **ciliary muscle** which regulates the curvature of the lens. The muscle fibres arise from the spur at the corneo-scleral junction and run backward into the ciliary body. Their contraction draws the ciliary body forward, relaxing the suspensory ligament which, in turn, enables the lens to become more convex and to focus objects on the retina for near vision. The ciliary muscle is supplied by parasympathetic fibres from the ciliary ganglion.

C. Prosections of the orbit

The eye lies in the orbit. With a skull, review the bones which form the walls of this bony cave (**6.8.8**). Note that the eye is protected anteriorly by the upper and lower eyelids, each of which comprise a cartilaginous **tarsal plate**, attached to margins of the orbit by the **orbital septum**, the **medial** and **lateral palpebral ligaments** (**6.8.9**). Superficial to these lies the facial muscle orbicularis oculi which contracts reflexly (blinking reflex) to close the eyelids in response to potentially dangerous stimuli such as sudden flashes of light, to stimulation of the cornea or

conjunctiva, or in sleep. Review the attachments of its different parts (p. 47).

Examine the fibroelastic membrane, **Tenon's capsule** (**6.8.10**), which partially encloses the

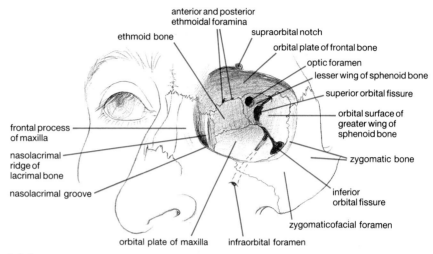

6.8.8
Bones of orbit.

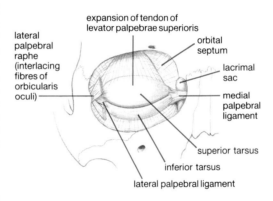

6.8.9
Eyelids and their attachment to orbital margin.

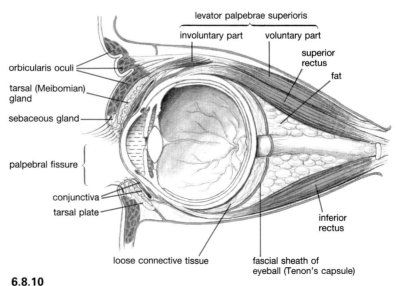

6.8.10
Sagittal section through orbit.

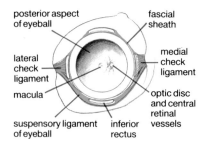

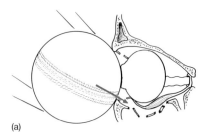

6.8.11
Fascia of orbit; suspensory ligament
of eyeball.

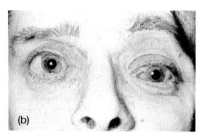

(a)

(b)

6.8.12
(a) Fracture of orbital floor and
rupture of suspensory ligament; (b)
resultant downward displacement of
left eye.

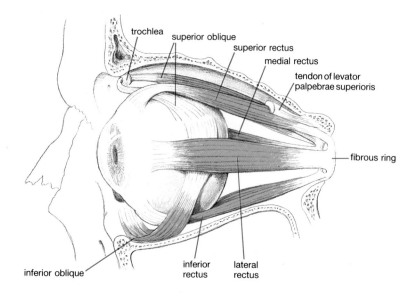

6.8.13
Extrinsic muscles of eyeball.

eyeball posteriorly and from which the eyeball may be surgically enucleated. The capsule is attached posteriorly to the optic nerve and anteriorly to the margin of the cornea, and is continuous with the fascial sheaths of the extrinsic ocular muscles as they pierce the capsule to insert into the sclera. From these sheaths, connective tissue expansions pass outward. Those associated with the medial and lateral recti form distinct triangular **medial** and **lateral check ligaments** which attach to the adjacent walls of the orbit and to the medial and lateral palpebral ligaments. The check ligaments may help to stabilize the horizontal position of the eyeball.

Inferior to the eyeball is a condensation of fascia, the **suspensory ligament** of the eye (**6.8.11**), which acts as a supporting sling for the eyeball and determines its vertical position. The suspensory ligament blends with the sheaths of inferior rectus and inferior oblique and, on either side, narrows to blend with the medial and lateral check ligaments thereby gaining attachment to the bony margins of the orbit. Trauma to the face which ruptures the suspensory ligament (**6.8.12a**), or damages its lateral attachment to a small tubercle on the inner aspect of the zygoma, causes downward displacement of the eyeball (**6.8.12b**).

Qu. 8I *What would be the consequences for vision if the attachment of the suspensory ligament were disrupted?*

Outside Tenon's capsule the eye is cushioned within pads of **orbital fat** which form a socket in which the eyeball moves. Part of this tissue is sensitive to immunoglobulins and in certain autoimmune diseases, particularly hyperthyroidism, it may hypertrophy, thrusting the eyeball forward to produce **exophthalmos** (**6.8.3**).

Extrinsic muscles of the eyeball
(6.8.13–6.8.15)

The eyeball can be moved within certain limits by the action of four **rectus muscles** and two **oblique muscles**. The recti take origin at the posterior aspect of the orbit from a fibrous ring which encircles the optic foramen and the medial part of the superior oblique fissure (**6.8.13**). The rectus muscles pass forward around the eyeball to insert into the sclera anterior to the horizontal axis of the globe but behind the corneo-scleral junction. Identify the **medial rectus** and the **lateral rectus** which rotate the eyeball as their names suggest (**6.8.15**). Identify also the **superior** and **inferior recti** which not only direct the gaze respectively upward or downward but also rotate the eyeball medially. The superior, medial, and inferior recti are supplied by the oculomotor nerve (III); the lateral rectus by the abducent nerve (VI).

Qu. 8J *How are the superior and inferior recti able to pull the eye medially?*

Locate the **superior oblique muscle** which arises from the medial wall of the orbit close to the fibrous ring. It passes forward above the medial rectus toward a fibrous pulley (trochlea) attached to the lacrimal bone. Here it becomes tendinous and the expanded tendon passes backward and laterally (beneath superior rectus) to insert into the posterolateral quadrant of the eyeball. Superior oblique rotates the cornea downward and laterally, enabling us to see the right or left lower corner of our field of vision.

Inferior oblique arises anteriorly from the floor of the orbit close to the lacrimal sac, and vertically beneath the trochlea. Its fibres extend backward and laterally beneath those of inferior rectus to insert into the posterolateral quadrant of the eyeball with superior oblique. The superior oblique is supplied by the trochlear nerve (IV), the inferior oblique by the oculomotor nerve (III).

Qu. 8K *In what directions does inferior oblique rotate the eyeball?*

The contraction of individual extraocular muscles would produce some rotation of the eyeball about an anteroposterior axis. Such rotation is *not* seen in normal movements of the eyeball which are the result of co-ordinated contractions and relaxations of all the extraocular muscles. If, however, one or more muscles is paralysed, then the action of the remaining muscles may cause rotation of the globe.

Locate **levator palpebrae** (**6.8.10**), the muscle which raises the upper eyelid. It arises from the posterior aspect of the roof of the orbit immediately above the origin of superior rectus and inserts into the skin and tarsal plate of the upper eyelid. Most of the muscle is striated and supplied by the oculomotor nerve (III), but some of the deep fibres are smooth, and supplied by sympathetic nerves.

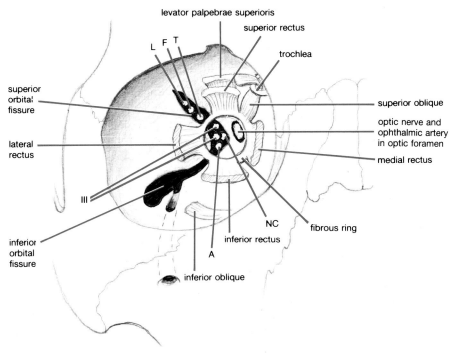

6.8.14
Structures at back of orbit. Lacrimal (L); frontal (F); trochlear (T); nasociliary (NC); and abducent (A) nerves.

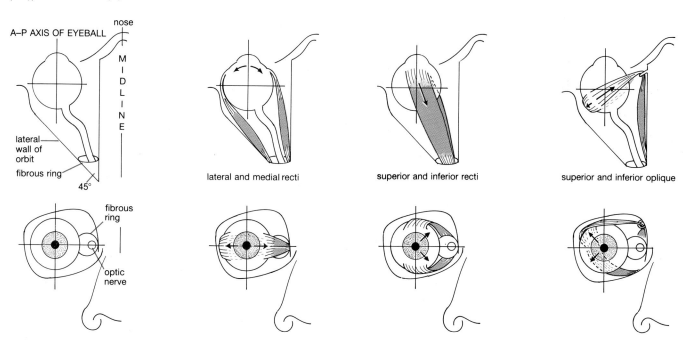

6.8.15
Movements of eyeball and extrinsic muscles.

Qu. 8L *Ptosis (drooping of the upper eyelid) occurs as a result of damage to either the oculomotor or sympathetic nerves. How would the ptosis differ in these two cases?*

Arteries and veins of the orbit (6.8.16)

Locate the **ophthalmic artery** which is a branch of the internal carotid artery given off as this vessel leaves the cavernous sinus. The ophthalmic artery passes through the optic foramen with the optic nerve and provides the **central artery of the retina** which enters the nerve to supply it and the retina. Follow the artery as it crosses over the optic nerve and passes anteriorly. Locate a **lacrimal artery** which passes along the lateral border of the orbit to the lacrimal gland and eyelids; long and short **ciliary arter-**

ies which pierce the eyeball around the optic nerve to enter the choroid coat; a large **supraorbital artery** which, with the supraorbital nerve, runs forward along the roof of the orbit to leave through the supraorbital notch; branches to the muscles of the orbit, and **ethmoid arteries** which leave the medial wall of the orbit to enter the ethmoidal air sinuses (the anterior ethmoid artery briefly re-enters the cranial cavity and then enters the nose through its roof).

The **ophthalmic veins** (superior and inferior) communicate with facial veins at the medial margin of the orbit, and with the pterygoid venous plexus and deep facial veins through the inferior orbital fissure. They receive tributaries from the contents of the orbit and pass posteriorly through the superior orbital fissure to drain into the cavernous sinus.

Nerves of the orbit (6.8.16, 6.8.17)

With the exception of the optic nerve, all nerves entering the orbit traverse the superior orbital fissure (see **6.8.14**).

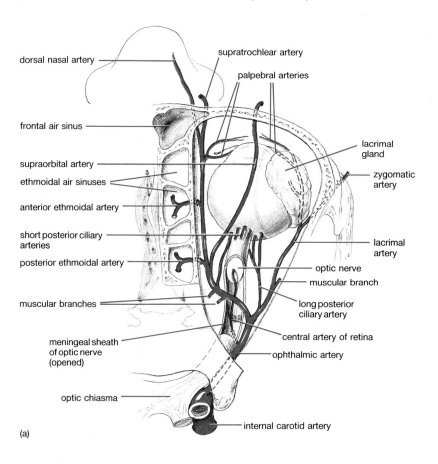

(a)

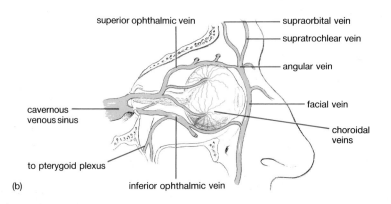

(b)

6.8.16
(a) Arteries and (b) veins of orbit.

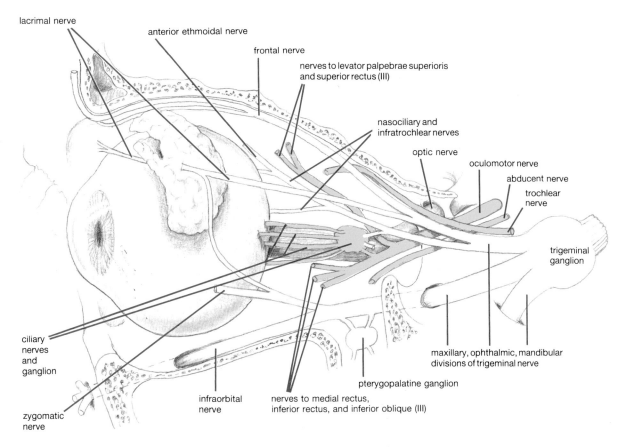

lacrimal nerve

anterior ethmoidal nerve

frontal nerve

nerves to levator palpebrae superioris
and superior rectus (III)

nasociliary and
infratrochlear nerves

optic nerve

oculomotor nerve

abducent nerve

trochlear
nerve

trigeminal
ganglion

ciliary
nerves
and
ganglion

zygomatic
nerve

infraorbital
nerve

nerves to medial rectus,
inferior rectus, and inferior oblique (III)

pterygopalatine ganglion

maxillary, ophthalmic, mandibular
divisions of trigeminal nerve

6.8.17
Nerves of orbit. Parasympathetic fibres to lacrimal gland (*see text).

Identify the **ophthalmic division of the trigeminal nerve** (V), and its branches which supply sensation to the eyeball and conjunctiva (long cillary nerves), the skin over the forehead (supraorbital and lacrimal branches) and nose (supratrochlear, infratrochlear, the external nasal branches), and the ethmoid sinuses and upper part of the nose (p. 51).

Identify the thin **trochlear nerve** as it passes through the superior orbital fissure just outside the fibrous ring to supply superior oblique.

Identify also the **abducent nerve** (VI) which passes through the superior orbital fissure within the fibrous ring to lie inside the muscular sheath and supply lateral rectus.

Identify the **oculomotor nerve** (III) which also lies within the muscular sheath, supplying all the other muscles of the orbit. Locate its branch to the superior rectus which passes upward through this muscle to supply levator palpebrae. The branch to the inferior oblique, in addition to somatic motor fibres, carries parasympathetic fibres which synapse in the **ciliary ganglion**. Search for this ganglion lying lateral to the optic nerve in the orbit, and for its short ciliary branches which pierce the sclera of the eyeball to supply the **sphincter pupillae** and the **ciliary muscle**.

Qu. 8M *If the oculomotor nerve (III) were damaged, what would be the consequences?*

Lacrimal apparatus (6.8.4)

To keep the conjunctiva and cornea moist and to wash away foreign particles or irritants, aqueous tear fluid is secreted by the lacrimal gland which is situated laterally under the orbital margin and upper eyelid. The secretions, which have antiseptic properties, pass into the conjunctival sac through a number of small ducts which pierce the upper fornix of the conjunctiva. Tear fluid is continuously passed across the surface of the eye by the process of blinking.

Qu. 8N *What, then, would be one serious consequence of damage to the facial nerve?*

Tear fluid drains by capillarity into the lacrimal puncta on the medial aspects of the eyelids, then through the elastic superior and inferior canaliculi to the lacrimal sac. During blinking, drainage may be facilitated by contraction of a few (lacrimal) fibres of orbicularis oculi which are attached behind the lacrimal sac. Tears drain from the lacrimal sac via the nasolacrimal duct.

Qu. 8O *To where do these lacrimal secretions eventually drain?*

Tears do not normally overflow on to the cheeks or dry rapidly because the lipid-rich secretions from the **tarsal glands** (Meibomian glands) increase the surface tension and form a film over the aqueous tears.

D. Radiology

Examine a P–A radiograph of the skull taken with the correct tilt to show the orbits (**6.8.18**). Note the superior orbital fissure (arrowhead) and the position of the optic foramen (arrow).

Examine the CTs of the orbit. **6.8.19** is a horizontal section of the orbit at the level of the optic nerve (O); identify the medial (m) and lateral (1) rectus muscles and note the thin, fragile ethmoid bone.

Examine radiograph **6.8.20**; contrast medium has been injected through very fine catheters (small open arrows) inserted into the inferior canaliculus of each eyelid. The contrast medium has filled the lacrimal sac on each side and has passed through the nasolacrimal duct (arrowhead) to spill on to the floor of the nasal cavity. A little reflux on to the conjunctiva has occurred through the superior canaliculus on each side (large open arrow).

Requirements:

Skull.

Eyeball of an ox (obtainable from butcher).

Prosections of the orbit with the orbital plate of the frontal bone removed to show levator palpebrae and its insertion into the upper tarsal plate; extrinsic muscles of the orbit; blood vessels and nerves of the orbit; the ciliary ganglion and its connections; anterior prosection of the orbit to show the suspensory ligament of the eyeball.

Radiographs and CTs of the orbit and lacrimal apparatus.

Ophthalmoscope.

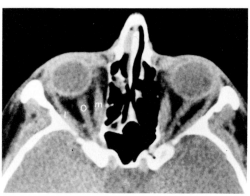

6.8.18
P–A radiograph of bones of orbit.

6.8.19
Horizontal CT through orbits.

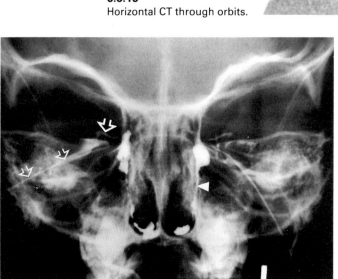

6.8.20
Nasolacrimal apparatus outlined by radio-opaque material.

Seminar 9

Ear

The **aim** of this seminar is to study the separate but functionally integrated components of the auditory apparatus, the vestibular apparatus which provides information about our position in and movement through space, and the vestibulocochlear nerve (VIII) which transmits the sensory information to the central nervous system. This is done by examination of the living, and study of prosections, dissected temporal bones, models and radiographs. It is important that the histology and neurophysiology of the ear is read in conjunction with your study of the structure.

Development of the ear

(6.9.1)

In the early embryo rudiments of the inner ear appear as bilateral thickenings of the ectoderm, the **otic placodes**, close to the developing hindbrain. These invaginate to form **otic pits** which separate from the surface to form **otic vesicles**. As each otic vesicle enlarges it becomes divided into an upper **vestibular** and a lower **cochlear** part, and, from its medial aspect, a vertical diverticulum elongates to form the **endolymphatic duct**. From the vestibular part arise three diverticula situated at right angles to each other; their middle parts coalesce and disappear to leave peripheral **semicircular canals** attached to a central **utricle**. The cochlear part differentiates to form a ventral spirally-coiled **cochlear duct** and a dorsal chamber, the **saccule**, which is connected to the cochlear duct by the narrow **ductus reuniens** and to the utricle by the **utriculosaccular canal**. All these derivatives of the otic vesicle which constitute the **membranous labyrinth** are surrounded by a mesenchymal capsule which becomes cartilaginous and then ossifies to form the **bony labyrinth**. The vestibulocochlear nerve appears first as a group of cells which are partly derived from the neural crest and partly from the wall of the otocyst; those which form the ganglion lie between the hindbrain and the otic vesicle. From these cells a peripheral process passes to the developing membranous labyrinth, and a central process into the hindbrain.

The middle ear and pharyngotympanic tube are formed largely from the dorsal part of the endoderm-lined 1st pharyngeal pouch (**tubotympanic recess**), with contributions from the 2nd and 3rd pouches. The 1st pharyngeal pouch surrounds the dorsal end of the cartilages of the 1st and 2nd pharyngeal arches which form the ear ossicles (the malleus and

incus from the 1st arch; the stapes from the 2nd arch). Unlike other cranial air sinuses, the mastoid antrum is well developed at birth.

The external auditory meatus forms from the 1st ectodermal cleft which extends inward towards the endodermal tubotympanic recess. At the junction the tympanic membrane is formed from ectoderm, endoderm, and intervening mesenchyme. The pinna develops from several ectoderm-covered mesodermal hillocks which form around the 1st ectodermal cleft. Of these the most prominent from the 1st pharyngeal arch forms the tragus.

A. Living anatomy

The entrance to the outer ear is surrounded by the elastic cartilaginous auricle or **pinna** (6.9.2). The pinna acts as a complicated diffraction grating which affects the transfer of particularly high-frequency sounds into the external auditory meatus. Its asymmetric shape and asymmetric position (with respect to the meatus) help in the localization of the source of a sound, particularly in distinguishing front from back and up from down.

Qu. 9A *When a sound we wish to hear is faint we often cup a hand behind an ear. How does this help, and what other simple manoeuvre is also useful?*

Note the component parts of the pinna, particularly the **tragus**. Feel the pinna to determine which parts are supported by cartilage, and note that the skin is firmly attached to the cartilage.

Qu. 9B *Why do severe blows to the pinna result in deformed 'cauliflower' ears?*

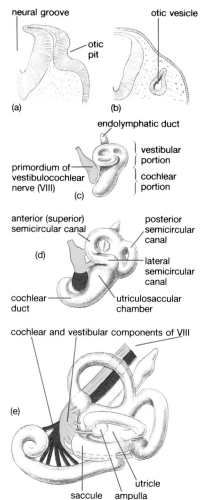

6.9.1
Development of internal ear:
(a,b) transverse sections; (c–e) lateral views.

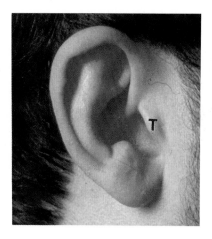

6.9.2
Pinna.Tragus (T).

Examine the **external auditory meatus** which passes inward from the pinna. The entrance to the meatus, like that of the nose, is protected (in older men) by coarse hairs (vibrissae), and the waxy secretions of **ceruminous glands** also trap incoming particles. Gently tug on the pinna and note that the outer part of the meatus moves with the pinna. This is because the outer one-third of the meatus has a fibrocartilaginous framework, whereas the inner two-thirds passes through the tympanic part of the temporal bone. The meatus is lined by skin which is firmly attached to the underlying cartilage or bone.

Qu. 9C   *Why is an infected spot or boil in the external auditory meatus more painful than one, for instance, on the face?*

Shine the light of an auriscope into the meatus and note that it is not a straight tube. Rather, it has a gentle sigmoid course as it passes medially. It slopes at first forward and slightly upward, then backward and upward, and finally backward and downward. In order to straighten the meatus for examination with an auriscope, the pinna must be pulled backward and upward. This is unnecessary in an infant because the external auditory meatus is much shorter, and the tympanic membrane therefore more superficial (compare a fetal (**6.9.3**) and adult skull). The bony ring around the tympanum, although it starts to ossify shortly after birth, grows slowly and does not develop fully until puberty. The superficial tympanic membrane in an infant is more easily visualized, but also more liable to damage, for example during attempts to remove foreign bodies from the ear.

Introduce the speculum of an auriscope into the external auditory meatus of your partner, pulling the auricle backward and upward. Note

the presence of any wax and, if possible, examine the **tympanic membrane** (**6.9.4**). The tympanic membrane is normally pale; a reddened membrane indicates infection of the middle ear which may perforate the drum if sufficiently severe. From its centre, a cone of light should be visible extending downward and forward to the floor of the meatus. The cone of light is caused by the shape of the tympanic membrane, a shallow cone of tense fibrous tissue, with its apex (**umbo**) directed into the middle ear, and set obliquely (55°) so that the floor of the meatus is longer than its roof. Passing upward and forward from the apex of the cone of light, your may be able to detect the handle of the malleus and, almost at its upper limit, the flaccid part of the tympanic membrane, separated from the tense part by malleolar folds. Posterior to the handle of the malleus, the tip of the long process of the incus is sometimes just visible through the membrane.

Demonstrate the continuity between the pharynx and middle ear by closing your mouth, pinching your nose, and blowing out. You should feel a sensation within your ears and hear a low-pitched sound as the tympanic membrane is forced outward by the increased pressure within the middle ear cavity transmitted from the nasopharynx via the pharyngotympanic tube.

For testing of auditory and vestibular function, see p. 119.

B. Prosections of the ear

External ear

Examine horizontal and vertical sections of the head that pass through the external auditory meatus. Note its sigmoid course, the extent of

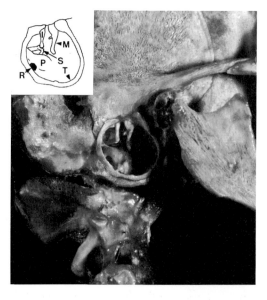

6.9.3
Fetal skull showing superficial position of tympanic ring and auditory ossicles.

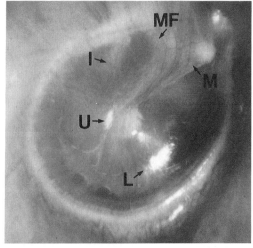

6.9.4
Tympanic membrane. Malleus (M); incus (I); stapes (S); promontory (P); and round window (R).

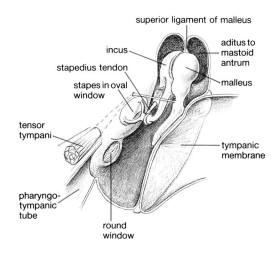

6.9.5
Anterior view of right middle ear cavity showing ossicles.

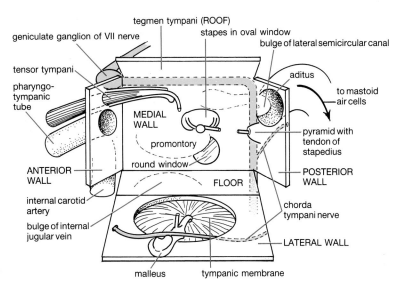

6.9.6
Walls of middle ear (diagrammatic, in form of an opened box).

its cartilaginous and bony portions, and the oblique orientation of the tympanic membrane which forms part of the lateral wall of the middle ear.

Middle ear (6.9.5–6.9.7)

Now examine prosections of the middle ear, dissected temporal bones, and a model (if available). Examine first the size and shape of the cavity of the middle ear (tympanic cavity) within the petrous temporal bone (**6.9.5**); it is a slim irregular vertical cavity, shaped roughly like a matchbox (**6.9.6**), which is oriented so that its anterior wall is more medial than its posterior and its roof more lateral than its floor. The cavity is wider in its upper one-third which lies above the level of the tympanic membrane, forming the **epitympanic recess**. The cavity and contents are lined by partly ciliated cuboidal epithelium.

Lateral wall. The tympanic membrane separates the external auditory meatus from the middle ear cavity, which extends upward above the membrane into the epitympanic recess. Note that the head of the malleus lies in the epitympanic recess, while its handle (covered with the mucous membrane which lines the middle ear) is attached to the upper part of the inner surface of the tympanic membrane. Between the tympanic and mucous membranes the chorda tympani branch of the facial nerve (VII) runs from the posterior to the anterior wall.

Posterior wall. In the upper part of the posterior wall identify an opening, the **aditus**, which connects the epitympanic recess with the **mastoid antrum**. The antrum is a single large mastoid air cell which connects with the other, smaller cells which honeycomb the mastoid process to a variable extent. Infection in the middle ear can spread to the mastoid air cells from which it may be difficult to eradicate if the air cells com-

municate poorly with each other. On the medial wall of the aditus note the bulge created by the lateral semicircular canal of the inner ear. Below the aditus is a small projection, the **pyramid**, from the apex of which the **stapedius** muscle emerges to attach to the neck of the stapes. Within the posterior wall the facial nerve descends to the stylomastoid foramen, gives off the chorda tympani, and supplies stapedius.

Anterior wall. On the narrow anterior wall locate the orifice of the **pharyngotympanic (auditory or Eustachian) tube** through which the middle ear communicates with the nasopharynx. The pharyngotympanic tube is 3–4 cm long; its upper one-third passes obliquely forward and medially through the temporal bone but its lower two-thirds has a cartilaginous roof and fibrous floor. Review the opening of the pharyngotympanic tube and tubal elevation posterior to the inferior meatus of the nose within the nasopharynx (see **6.5.1**).

Qu. 9D *Often during descent in an aircraft one's ears feel as if they might 'burst'. Why is this and how can the sensation be relieved?*

Immediately above the orifice of the pharyngotympanic tube and separated from it by a thin bony septum locate a smaller orifice from which **tensor tympani** emerges; the muscle passes backward along the medial wall before bending laterally around a minute bony pulley to attach to the upper part of the handle of the malleus.

Qu. 9E *What are the functions of tensor tympani and stapedius?*

Medial wall. The medial wall of the tympanic cavity abuts on the inner ear to which it is connected by two 'windows'. Locate the **oval window** (fenestra ovalis) in which is attached the baseplate of the stapes and, immediately below it, the smaller **round window** which is closed by a fibrous membrane (**6.9.6**). Immediately

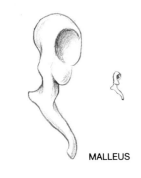

MALLEUS

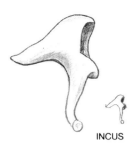

INCUS

STAPES

6.9.7
Ossicles of middle ear.

anterior to these openings and projecting into the middle ear is a rounded bony eminence, the **promontory**, which overlies the first turn of the cochlea. Above the promontory, the facial nerve runs backward through the medial wall before turning downward beneath the aditus to the mastoid antrum.

Examine the **roof** and **floor** of the tympanic cavity. The roof (tegmen tympani) is very thin and lies immediately below the temporal lobe of the brain. Examine the base of a skull and CT (**6.9.9**) and note that the narrow floor is also thin and separates the tympanic cavity from the jugular fossa and carotid canal. Through the floor tympanic branches of the glossopharyngeal nerve and sympathetic fibres from the carotid plexus reach the tympanic cavity.

Qu. 9F *A fracture through the base of the skull often results in bleeding from the external auditory meatus. Why is this?*

Qu. 9G *If an infection of the middle ear were allowed to continue, to where might the infection spread?*

Lying within the middle ear cavity are the auditory ossicles (**6.9.5, 6.9.7**), a chain of articulated bones which transmit vibration from the tympanic membrane to the oval window and inner ear. Examine the ossicles in a prosection of the middle ear and as isolated bones. The hammer-shaped **malleus** has a downward- and backward-projecting handle which is attached to the tympanic membrane down to the umbo (**6.9.4, 6.9.5**). Its **head** lies within the epitympanic recess and is attached by fine ligaments to its roof and walls. Immediately beneath the neck, the small **lateral process** of the malleus projects into the tympanic membrane and creates 'folds' (**anterior and posterior malleolar folds**) which separate the upper flaccid part of the membrane from its lower 'tense' part; the **anterior process** of the malleus is attached by a

ligament (both derived from Meckel's cartilage) to the anterior wall of the recess. Tensor tympani is also attached just below the head of the malleus. On the posterior aspect of the head is the saddle-shaped synovial articulation with the **incus** (anvil). Examine this minute bone: its **head** articulates with the malleus; its **short process** projects posteriorly to be anchored by a ligament to the wall of the epitympanic recess; and its **long process** runs downward and backward parallel with the handle of the malleus to articulate at its tip with the **stapes** by a tiny ball-and-socket synovial joint. Examine the stirrup-shaped **stapes**: its **head** articulates with the long process of the incus, while its **neck** receives the attachment of stapedius and its two curved limbs join the oval **base** which fills the oval window and is attached to its margins by a ring of ligamentous fibres.

Inner ear (6.9.8)

The inner ear, which lies within the petrous temporal bone, consists of a **bony labyrinth** (lined with endosteum) within which lies a **membranous labyrinth**. It forms two quite distinct functional units, the **cochlea** which is the organ of hearing and the **vestibule** by which we perceive our orientation in and movement through space. The bony labyrinth is filled with **perilymph** and is in continuity with the CSF in the subarachnoid space through the **cochlear canaliculus** (perilymphatic duct) which opens on to the inferior aspect of the petrous temporal bone. Suspended within the bony labyrinth, but attached in part to the bone, is the membranous labyrinth (i.e. the cochlear canal and vestibule) which is filled with **endolymph**, a fluid which is secreted from the vascular lateral wall of the cochlear canal into the membranous labyrinth; it has an ionic composition very different to that of the perilymph.

Qu. 9H *Endolymph resembles intracellular fluid in its composition. What then are the relative concentrations of K and Na ions?*

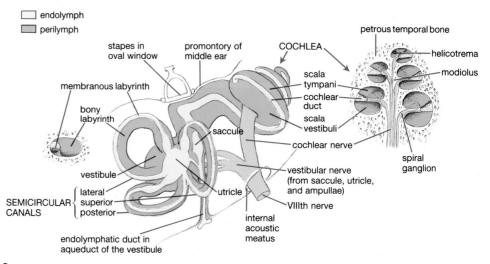

6.9.8
Inner ear.

The bony **cochlear canal** is spiral in shape and twisted 2.5 times around a central pillar, the **modiolus**, through which the cochlear nerve (VIII) is distributed to the organ of hearing. Its base is directed towards the **internal acoustic meatus** while its apex points forward and laterally. The membranous endolymph-filled **cochlear duct (scala media)** is wedge-shaped and attached by its apex to the modiolus and by its base to the lateral wall of the bony labyrinth thus dividing the perilymph-filled cavity into two channels (**scala vestibuli** and **scala tympani**) which are continuous with one another at the tip of the spiral cochlea (helicotrema). The roof and floor of the cochlear duct (scala media) therefore comprise two free membranes, respectively, the delicate **vestibular membrane** and the highly specialized **basilar membrane**.

The stapes, which is attached by its base within the oval window, transmits vibrations of the ossicles into the perilymph of the scala vestibuli and scala tympani, the lower end of which impinges on the **round window**. The round window is closed by the **secondary tympanic membrane**; through it, pressure waves transmitted to the inner ear through the oval window equilibrate with the middle ear.

Vibrations of the ossicles transmitted through the oval window to the perilymph cause vibration of the endolymph and basilar membrane, on which lies the organ of hearing, the **spiral organ of Corti**. This comprises an inner and outer row of hair cells (and various supporting cells) which move with the basilar membrane. The tips of the 'hairs' (cilia) of the outer hair cells are embedded in the **tectorial membrane** (an elongated tongue of soft collagenous material projecting from the shelf of bone) and are distorted by vibrations of the basilar membrane. The vibrations are transduced into receptor potentials via the cilia and these potentials contribute to internal electromechanical regulation within the cochlea. By contrast, the cilia of the inner hair cells are not firmly embedded in the tectorial membrane, but respond to viscous forces from the moving cochlear fluid. The receptor potentials of *inner* hair cells provide at least 95 per cent of the auditory signals that are transmitted to the central nervous system; they synaptically excite afferent terminals of the **cochlear (VIII) nerve**. The cell bodies of these sensory nerves form the **spiral ganglion** within the modiolus.

The vestibulocochlear (VIII) nerve also contains efferent fibres, which either synapse on to the cochlear hair cells or make axo–axonic contact with afferent fibres, and which modify their output.

The **vestibular system** (6.9.8) comprises three **semicircular canals** which detect acceleration, and the **utricle** and **saccule** which detect orientation. Like the cochlea it consists of a complex bony labyrinth filled with perilymph within which is attached an endolymph-filled membranous canal. Identify the three semicircular canals and examine their orientation in three orthogonal planes. The **lateral** semicircular canal lies horizontally and projects laterally into the aditus to the mastoid antrum. The **superior** and **posterior** semicircular canals both lie vertically above it and share a medially-placed common canal. They are orientated like an open book facing laterally and standing vertically on a shelf. Note that the superior canal of one side is aligned parallel to the posterior canal of the other side. The superior canal usually produces a ridge which can be located on the roof of the petrous temporal bone. Laterally, where they enter the utricle, the semicircular canals are enlarged to form **ampullae** in which are located the sensory receptors, specialized hair cells the tips of which are embedded in a gelatinous **cupula**. Displacement of the endolymph causes bending of the cilia which is transduced and transmitted to terminals of the vestibular (VIII) nerve. Each canal therefore responds to acceleration (rotation) in its own plane.

Similar sensory receptors comprising hair cells, with cilia embedded in otolith-containing gelatinous material (**maculae**), are found in the saccule and utricle. The macula of the saccule lies in a vertical plane roughly perpendicular to that in the utricle. Both are responsive to positional changes of the body. Gravity pulls down the otolith-containing material; this bends the cilia of the hair cells which transduce and transmit the information to sensory axons of the vestibular (VIII) nerve, cell bodies of which lie in the vestibular ganglion situated in the internal auditory meatus. Efferent fibres, analogous in both their synaptic contact and function to those to the cochlea, are found in the vestibular (VIII) nerve.

The cochlear and vestibular parts of the membranous labyrinth are continuous via the **ductus reuniens** and together form a closed system from which the endolymph eventually passes into the circulation through the **endolymphatic duct**, a blind-ending sac which arises from both the utricle and saccule and impinges on the dura mater on the posterior aspect of the petrous temporal bone. It is surrounded by well vascularized connective tissue through which endolymph is resorbed into the bloodstream.

Blood supply to the ear

Note the blood supply to different parts of the ear. The external and middle ear are supplied from the external carotid system; the inner ear from the vertebral.

The pinna is supplied by superficial temporal and posterior auricular arteries, and the external meatus by small (deep auricular) branches of the maxillary artery. The middle ear derives blood from the maxillary (anterior tympanic) and posterior auricular arteries which form a ring around the tympanic membrane. The inner ear is supplied by the labyrinthine artery which usually arises from the basilar artery.

Innervation of the ear

The innervation of the external ear reflects its development at the junction of the first and more caudal pharyngeal arches. The tragus and upper part of the pinna is supplied by the auriculotemporal nerve (mandibular V), the remainder by the greater auricular and lesser occipital nerves (C2, C3). The anterior part of the external auditory meatus and tympanic membrane is supplied by the auriculotemporal nerve and its posterior part by the auricular branch of the vagus nerve (X). A few fibres of the facial nerve (VII) also supply the tympanic membrane.

Qu. 9I *Banqueting Romans used various devices (water, ivory probes) to stimulate the external auditory meatus. Why might that be?*

The middle ear and pharyngotympanic tube are supplied by the glossopharyngeal (IX) nerve through its tympanic branch. The mastoid air cells are supplied by the nervus spinosus branch of the mandibular nerve (V) (p. 121).

The vestibular and cochlear fibres of the VIII nerve pass from the ganglion cells of the inner ear through the internal auditory meatus, and across the cerebello-pontine angle to vestibular and cochlear nuclei lying in the medullary part of the brainstem.

The facial nerve (VII), containing motor, sensory, and secretomotor fibres, enters the internal acoustic meatus with the vestibulocochlear nerve and, on reaching the anterior part of the middle ear where the cell bodies of its sensory component form the geniculate ganglion, turns backward across the medial wall of the middle ear. On reaching the posterior wall of the middle ear, the nerve makes a second right-angled turn to descend through the posterior wall in

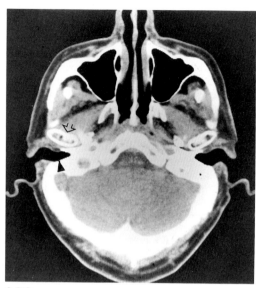

6.9.9
Horizontal CT through external auditory meatus.

the stylomastoid canal. Within the middle ear it supplies stapedius and gives off the chorda tympani which runs across the inner aspect of the tympanic membrane and malleus to exit through the anterior wall.

C. Radiology

Examine the horizontal CT through the petrous temporal bone (**6.9.9**); note the shape of the external auditory meatus (arrowhead) and the head of the mandible (open arrow) immediately in front of the meatus. Examine the oblique CT (**6.9.10**) through the petrous temporal bone. **6.9.10a** shows the middle ear cavity (M), mastoid air cells (A), thin tegmen tympani (T), first turn of the cochlea (C) forming the promontory (P), and the jugular bulb (J); **6.9.10b** shows the lateral semicircular canal (S) and internal auditory meatus (I). Examine **6.9.11**, a coronal CT through the middle ear cavity (m), tegmen tympani (T), and head of the malleus (h) in the epitympanic recess (e); note the canal for the

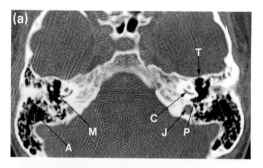

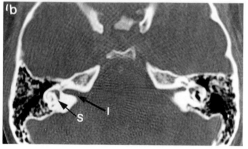

6.9.10
CT through middle and inner ears (see text).

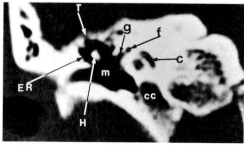

6.9.11
CT through middle ear.

facial nerve (f) and its geniculate ganglion (g) on the medial wall of the middle ear and the carotid canal (cc) which is separated from the floor of the middle ear by a very thin plate of bone. For other CTs through the ear see **6.12.9**.

Requirements:

Skull and removable skull cap; separate temporal bone with tegmen tympani removed; separate ossicles.

Prosections of the pinna and auricular muscles; horizontal section through the head to show the external and middle ears; the course of the vestibular and auditory parts of VIII nerve and the facial nerve passing through the middle ear; temporal bones to show the cavity and ossicles of the middle ear, the cochlea, and vestibular systems.

Model of the external, middle, and inner ears (if available). Radiographs and CTs of the temporal bone.

Auriscope.

Seminar 10

Blood supply and lymphatic drainage of the head and neck

The **aim** of this seminar is to study the blood supply and lymphatic drainage of the head and neck, in the living, and by means of prosections and radiographs.

Development

The arterial supply to the developing brain is derived initially from the paired dorsal aortae, which are progressively connected to the truncus arteriosus by a series of pharyngeal arch arteries. Of these, the 3rd arch artery persists to form the common carotid artery and stem of the internal carotid artery, the remainder of the internal carotid artery being formed from the dorsal aorta. The first two pharyngeal arch arteries largely disappear, but the distal portion of the 2nd persists as a tympanic branch of the carotid artery to the middle ear. Along the ventral aspect of the hindbrain a plexus of vessels develops which connects cranially with the internal carotid artery, and caudally with the vertebral artery. The face develops relatively late, and is supplied by a branch which grows out from the 3rd aortic arch to form the external carotid artery and its branches.

The anterior cardinal veins drain the head and brain. These form the internal jugular veins and, caudal to their junction with the developing subclavian veins, the brachiocephalic veins. An anastomosis between the two anterior cardinal veins carries blood from the left side to the right anterior cardinal vein and heart, forming the transverse part of the left brachiocephalic vein. The more caudal part of the left anterior cardinal vein regresses. (See Vol. 2, Chapter 5, Seminar 4.)

A. Living anatomy (6.10.1)

The head and neck is supplied with blood from two sources, the **common carotid arteries** and the **subclavian arteries**. Face forward in a normal, relaxed position and press downward behind the midpoint of the clavicle to feel the pulsations of the **subclavian artery** as it passes over the 1st rib to the axilla. Now ask your partner to face away from you and palpate gently backward on either side of the thyroid cartilage to feel the pulsations of the bifurcation of

the **common carotid artery** on each side. Now, with the head thrown back, palpate the same area.

Qu. 10A *What structure(s) now prevents you from feeling the pulsations?*

Two other arteries in the head and neck can be felt easily; both are branches of the external carotid artery. Clench your jaws and define the anterior border of masseter, then relax and palpate the lower border of the mandible immediately in front of the insertion of masseter to feel the **facial artery** as it crosses the jaw to reach the face. Next palpate the zygomatic process anterior to the ear to feel the **superficial temporal artery** which passes upward over the zygomatic process to supply the scalp.

The **external jugular vein** is often visible as it passes downward across sternomastoid muscle to enter the subclavian vein in the root of the

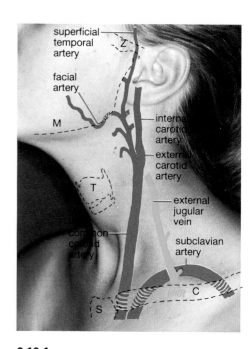

6.10.1

Surface markings of major vessels of the neck. Zygomatic arch (Z); mandible (M); thyroid cartilage (T); clavicle (C); sternum (S).

neck. If not, it can easily be made visible by the Valsalva manoeuvre in which one attempts to exhale against a closed glottis, thus raising intrathoracic pressure which is transmitted to the great veins. In thin people in the upright position, apparent pulsations of the internal jugular vein may be seen but, unless venous pressure is substantially raised (e.g. by right heart failure), these are pulsations transmitted from the underlying common carotid artery. Ask your partner to lie horizontal, or nearly so. True jugular venous pulsations (which result from contractions of the right atrium) can be seen in a normal subject in this position. Now ask your partner to sit upright and notice how the veins of the neck empty.

Stand behind your partner and gently try to feel for lymph nodes beneath the mandible, behind the angle of the mandible, behind the mastoid process, and along the occipital attachment of trapezius. If your partner has recently suffered an inflamed throat, tender enlarged nodes may be palpable just behind the angle of the mandible (upper deep cervical group of nodes). Finally, feel along the course of the common carotid artery for any nodes of the deep cervical chain, the lowest members of which are relatively inaccessible behind the medial aspect of the clavicle.

Retinal vessels can be viewed by use of an ophthalmoscope (see **6.8.5**). This is the only place in the body (apart from capillaries in the nail bed) where vessels can be directly visualized, and ophthalmoscopy is useful in detecting changes in vessels which result, for example, from hypertension.

B. Prosections of the arteries of the head and neck

The right common carotid artery arises from the brachiocephalic artery while the left arises directly from the arch of the aorta. Each common carotid divides into an **external carotid artery** which, through its branches, supplies the face, neck, and front and sides of the scalp, and an **internal carotid artery** which enters the skull to supply the brain and the orbit.

Each **subclavian artery** supplies branches to the root of the neck and the **vertebral artery**, which runs upward through the transverse processes of the cervical vertebrae to enter the skull and supply the brain.

The internal carotid and vertebral arteries anastomose inside the skull to form the arterial **circle of Willis** which lies at the base of the brain. Extracranial anastomoses between the internal and external carotid supplies and the vertebral arteries are extensive.

Subclavian artery

On a prosection of the root of the neck (**6.10.2**) examine the **right subclavian artery** which arises with the right common carotid artery from the brachiocephalic artery, a major branch of the arch of the aorta. Locate the **left subclavian artery** which arises directly from the arch of the aorta and note that, because the arch of the aorta is directed backward, its root lies posterior to that of the common carotid. Scalenus

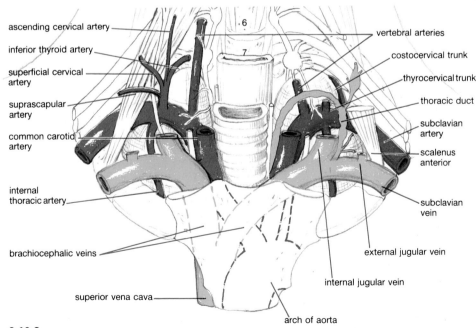

ascending cervical artery

inferior thyroid artery

superficial cervical artery

suprascapular artery

common carotid artery

internal thoracic artery

brachiocephalic veins

superior vena cava

arch of aorta

vertebral arteries

costocervical trunk

thyrocervical trunk

thoracic duct

subclavian artery

scalenus anterior

subclavian vein

external jugular vein

internal jugular vein

6.10.2
Root of neck.

anterior crosses the subclavian artery as it lies on the 1st rib, thereby dividing the artery (for descriptive purposes) into three parts which lie proximal (1st part), posterior (2nd part), and distal (3rd part) to the muscle.

Identify the three branches of the **1st part of the subclavian artery**. The **internal mammary artery** arises from the inferior aspect of the subclavian and passes downward into the thorax behind the costal cartilages (Vol. 2, Chapter 5, Seminar 2). The **vertebral artery** passes upward and backward to enter the foramen transversarium of C6 (see below). The **thyrocervical trunk** is a short vessel which passes upward and soon divides into the inferior thyroid, suprascapular, and transverse cervical arteries. Trace the **inferior thyroid artery** (see **6.6.14**) as it passes medially behind the carotid sheath to supply the inferior pole of the thyroid gland and adjacent structures including the larynx, trachea, oesophagus, and spinal cord; it anastomoses with branches of the vertebral artery and with the ascending pharyngeal, occipital, and deep cervical branches of the external carotid artery.

Qu. 10B *With which nerve are branches of the inferior thyroid artery closely related?*

Trace the **suprascapular artery** as it crosses the posterior triangle to reach the suprascapular

notch and supply muscles arising from the supra- and infraspinous fossae of the scapula (see Vol. 1, Chapter 6, Seminar 3). The **superficial cervical artery** also crosses the posterior triangle, lying parallel to but at a higher level than the suprascapular. It supplies the muscles of the neck and anastomoses with the occipital branch of the external carotid artery.

The **2nd part of the subclavian artery** gives off the **costocervical trunk** which arches posteriorly over the apex of the lung to the neck of the 1st rib, where it divides into the **superior intercostal artery** which enters the thorax to supply upper intercostal spaces and the **deep cervical artery** which ascends in the deep muscles of the neck, supplying them and anastomosing with the occipital and vertebral arteries.

The **3rd part of the subclavian artery** gives rise to the **dorsal scapular artery** which passes posteriorly across the root of the neck; follow it as it crosses the trunks of the brachial plexus to follow levator scapulae and take part in an anastomosis around the scapula.

Identify the **vertebral artery** (**6.10.3**) as it passes upward to the transverse process of the 6th cervical vertebra which lies at the apex of a triangle formed by scalenus anterior laterally and longus cervicis medially. The vertebral artery enters the foramen transversarium of C6 and, as it continues upward through the foramina transversaria, gives branches which run along the spinal nerve roots to reinforce the arterial supply of the spinal cord. The transverse process of the atlas extends further laterally than that of other cervical vertebrae so that the vertebral artery passes laterally from the foramen transversarium of the axis to enter that of the atlas. It then curves backward and medially grooving the upper aspect of the posterior arch and lateral mass of the atlas and pierces the posterior atlanto-occipital membrane between the atlas and axis to gain the vertebral canal. On entering the vertebral canal it pierces the dura mater and passes upward through the foramen magnum to join the opposite vertebral artery and form the **basilar artery** on the anterior aspect of the brainstem. Identify the branches of the vertebral artery within the skull (**6.10.4**): the **anterior spinal artery** joins its fellow of the opposite side to run downward in the anteromedian fissure of the cord; the **posterior inferior cerebellar artery** supplies not only part of the cerebellum but also the lateral part of the medulla of the brainstem. The two **posterior spinal arteries** which pass down the spinal cord medial to the dorsal roots may arise from either the vertebral or posterior inferior cerebellar arteries.

The **basilar artery** passes upward and forward on the ventral surface of the pons. It supplies the brainstem and gives **anterior inferior cerebellar** and **superior cerebellar** arteries which ramify beneath the tentorium to supply the cerebellum. Locate the **labyrinthine** artery to the inner ear; it arises from the superior cerebellar or basilar artery and runs across the cerebello-pontine angle with the facial and vestibulocochlear nerves to enter the internal audi-

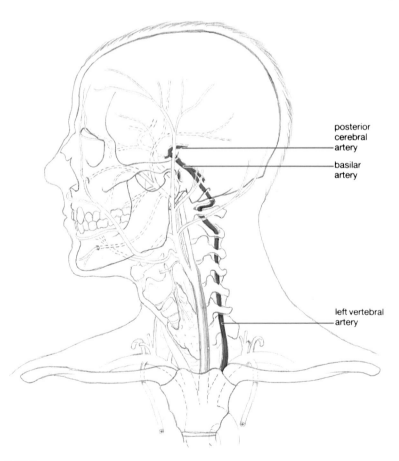

posterior
cerebral
artery

basilar
artery

left vertebral
artery

6.10.3
Vertebral artery and branches.

tory meatus. At the upper border of the pons the basilar artery ends by dividing into the two **posterior cerebral arteries** which pass around the brainstem above the tentorium to supply much of the inferior aspect of the temporal and occipital lobes of the brain, including the visual cortex. From them, small **posterior communicating arteries** pass forward to anastomose with the internal carotid system in the circle of Willis.

Qu. 10C *The vertebral supply may become restricted either by vascular disease or by arthritic projections from cervical vertebrae. What symptoms might occur when the subject twists his/her neck rapidly?*

Common carotid artery (6.10.5)

On a prosection which exposes the upper part of the thorax, locate the **right common carotid artery** as it arises from the brachiocephalic artery behind the right sternoclavicular joint, and the **left common carotid** which arises directly from the arch of the aorta, posterior and lateral to the brachiocephalic artery. Trace the common carotid arteries upward through the neck and note that they are enclosed within a tough, protective fibrous sheath — **the carotid sheath** — which also contains the internal jugular vein laterally and the vagus nerve posteriorly. This neurovascular bundle lies in the angle between the cervical transverse processes and their attached muscles posteriorly and the trachea, oesophagus, and their related structures medially. The common carotid arteries terminate, without giving other branches, at the upper border of the thyroid cartilage where they bifurcate to form the **external** and **internal carotid arteries**. At the bifurcation the vessels are more bulbous due to the presence of the baro- and chemoreceptors of the **carotid sinus** and **carotid bodies**.

Internal carotid artery (6.10.5)

Now trace the upward course within the carotid sheath of the internal carotid artery which, like the common carotid artery, has no branches in the neck. As it ascends it passes deep to the styloid process and its attached muscles to enter the **carotid canal** in the petrous temporal bone at the base of the skull (p. 29 and see **6.10.14**). Pass a probe into the carotid canal of an isolated skull and demonstrate that it passes forward and medially to emerge from the canal and into the foramen lacerum. From the foramen lacerum the artery enters the **cavernous sinus** where it lies against the side of the body of the sphenoid bone. It passes forward through the sinus, and then curves upward then backward on itself under cover of the anterior clinoid process, where it pierces the dura mater to enter the subarachnoid space. This very curved course is sometimes called the 'carotid syphon' and is well seen on angiograms (see **6.10.11**).

During its course through the petrosal temporal bone the internal cartoid artery gives small

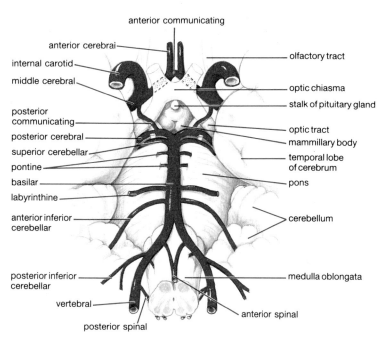

anterior communicating

anterior cerebrai

internal carotid

middle cerebral

posterior communicating

posterior cerebral

superior cerebellar

pontine

basilar

labyrinthine

anterior inferior cerebellar

posterior inferior cerebellar

vertebral

posterior spinal

olfactory tract

optic chiasma

stalk of pituitary gland

optic tract

mammillary body

temporal lobe of cerebrum

pons

cerebellum

medulla oblongata

anterior spinal

6.10.4
Intracranial branches of vertebral artery; circle of Willis.

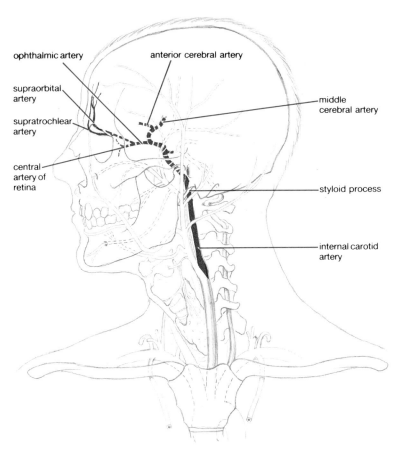

ophthalmic artery

supraorbital artery

supratrochlear artery

central artery of retina

anterior cerebral artery

middle cerebral artery

styloid process

internal carotid artery

6.10.5
Common and internal carotid arteries and branches.

branches to the middle ear and to the posterior part of the pituitary gland (**inferior hypophysial artery**). At the anterior clinoid process it gives off the **ophthalmic artery** whch passes through the optic canal beneath the optic nerve (II) to enter the orbit. It gives the **central artery of the retina**, an 'end artery' which enters and supplies the optic nerve and the retina. Within a prosected orbit identify the ophthalmic artery as it crosses obliquely from lateral to medial over the optic nerve giving branches to the contents of the orbit. These include a **lacrimal** branch to the gland and lateral part of the upper eyelid, muscular branches to the extraocular muscles, long and short **ciliary** arteries which pierce the globe to supply the choroid coat and iris, **ethmoidal** branches to the ethmoid air cells, and supraorbital, supratrochlear, and nasal branches which emerge from the orbit to supply the forehead and nose.

Qu. 10D *What would be the result of occlusion of the ophthalmic artery?*

Within the subarachnoid space identify the two terminal branches of the internal carotid artery, the **middle cerebral artery (6.10.4)** which supplies most of the lateral aspect of the brain, and the **anterior cerebral artery** which supplies the medial aspect and the uppermost part of the lateral aspect of the brain. An **anterior communicating artery** connects the two anterior

cerebral vessels and this, with the posterior communicating artery, completes the circle of Willis and the anastomosis between the vertebral and carotid systems.

Qu. 10E *If the internal carotid artery became narrowed by vascular disease, what symptoms would you expect?*

External carotid artery (6.10.6)

Identify the **external carotid artery** on a prosection and trace its course. It originates from the common carotid artery opposite the upper border of the thyroid cartilage, where it lies anterior to the internal carotid artery, and passes upward and anterior (superficial) to the styloid process to become enveloped by the parotid gland. At the neck of the mandible it divides into its two terminal branches, the **maxillary artery** and the **superficial temporal artery**. Since neither the common nor the internal carotid artery supply branches to the head, the external carotid arteries have many branches which you should identify, starting at the origin of the artery.

The **superior thyroid artery** loops downward to supply the upper pole of the thyroid gland and gives branches to adjacent parts of the larynx, in particular a superior laryngeal branch which enters the larynx through the thyrohyoid membrane with the internal laryngeal nerve.

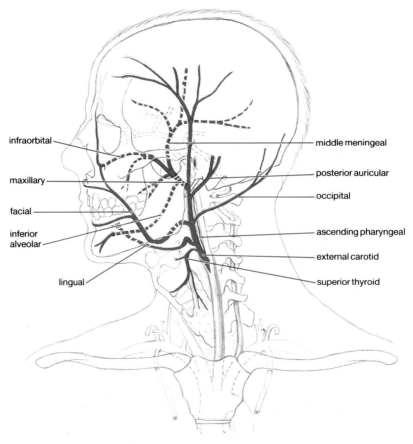

6.10.6
External carotid artery and branches.

The slender **ascending pharyngeal artery** ascends to the base of the skull on the side of the pharynx which it supplies, giving branches to the tonsil, the middle ear, and, at its termination, small meningeal vessels to the dura.

The **lingual artery** provides the principal supply of the tongue and floor of mouth; it arises opposite the tip of the greater horn of the hyoid bone and forms a small upward loop before passing anteriorly and deep to the hyoglossus muscle to enter the tongue (see **6.4.6**).

Qu. 10F *For what reason do many of the branches of the external carotid artery have a looped course?*

The **facial artery** (**6.10.6**, see **6.3.6**) arises close to, and commonly with, the lingual artery. Trace its course as it loops upward on the side of the pharynx, passing deep to the posterior belly of digastric and the submandibular salivary gland. It curves downward over the posterior pole of the submandibular gland to reach the lower border of the mandible immediately in front of the masseter. Follow it as it enters the face and passes obliquely upward to the angle of the mouth and thence along the side of the nose to the medial angle of the eye. Within the face it has a tortuous course to accommodate the movements of the face and jaw. At its origin it gives off the ascending palatine artery which supplies the pharynx and the soft palate. A **tonsillar artery** usually arises directly from the apex of the loop of the facial artery (which lies immediately lateral to the tonsil and is at risk in tonsillectomy). Before crossing the mandible the facial artery supplies the submandibular salivary gland and gives a relatively large branch, the **submental artery** which runs forwards on mylohyoid to anastomose with branches of the lingual and inferior alveolar arteries. As the facial artery runs upward on the face it gives many branches, including those to the lower and upper lips (inferior and superior labial branches), which anastomose freely with those of the opposite side, to the nose, and to the lacrimal sac.

Qu. 10G *If a labial artery were cut on one side, how might the bleeding be stopped?*

The external carotid artery gives two posteriorly-directed branches, the **occipital** and **posterior auricular arteries**. Trace the occipital artery as it runs upward under sternomastoid to supply the back of the scalp; and the posterior auricular artery which arises a little higher, and supplies the posterior surface of the ear.

The **superficial temporal artery** (**6.10.6**), the smaller terminal branch of the external carotid, begins in the parotid gland behind the neck of the mandible. Locate it as it passes upward, crossing the root of the zygomatic arch immediately in front of the ear. It supplies branches to the ear and face, including the **transverse facial artery** which runs forward parallel with the parotid duct and then enters the temporal region to supply the scalp.

The arteries that supply the scalp anastomose freely, as do branches of the external carotid on both sides of the face and branches of the internal carotid (e.g. the supratrochlear and supraorbital branches) so that facial wounds can continue to bleed if an external carotid has been damaged and ligated on one side and a clamp has been placed on the other.

The **maxillary artery** (**6.10.6**) is the larger of the two terminal divisions of the external carotid artery. It originates in the parotid gland behind the neck of the mandible, passes forward into the infratemporal fossa where it gives numerous branches, and finally passes through the pterygomaxillary fissure into the pterygopalatine fossa where its terminal branches supply the nose, nasopharynx and palate. It can conveniently be considered as having three groups of branches:

• The first group, in general, follows branches of the mandibular (Vth) nerve; small branches supply the external auditory meatus (deep auricular artery), the tympanic membrane (anterior tympanic artery) and the temporomandibular joint, while two larger branches supply the meninges (middle and accessory meningeal arteries) and the teeth and gums of the lower jaw (inferior alveolar artery).

Locate the **middle meningeal artery** as it arises from the maxillary artery in the infratemporal fossa and passes upward to enter the skull through the foramen spinosum. Within the skull it lies in a groove on the squamous part of the temporal bone; here it divides into two branches which sweep forward and then backward to supply a large area of cranium and its dura mater. The vessels lie in the extradural space and are held tightly to the bone by fibrous tissue. If the bone (which is often thin in the region of the pterion) is fractured, the artery may be torn. The fibrous connections prevent the artery from constricting adequately so that blood may leak slowly into the extradural space and cause increased intracranial pressure.

The **inferior alveolar artery** accompanies the inferior alveolar nerve to the medial aspect of the lingula of the mandible where both enter the mandibular canal to supply the bone, teeth and gums of the lower jaw. A branch emerges through the mental foramen to supply the skin over the chin. Immediately before the artery enters the mandibular canal a branch is given off (mylohyoid artery) which runs forward superficial to mylohyoid to supply the surrounding muscles.

• The second group of branches supplies the muscles of mastication and buccinator.

• The third group of branches supply structures originating from the maxillary process (and thus supplied by the maxillary nerve (V)), namely the bone, teeth, and gums of the upper jaw, the palate, nose, paranasal sinuses, and nasopharynx. As the maxillary artery passes into the pterygopalatine fossa it divides into several terminal branches. Locate the posterior superior alveolar artery which descends on the infratemporal surface of the maxilla to supply the molar and premolar teeth, gums, and maxillary sinus. The infraorbital artery enters the infraorbital fissure and canal and leaves through the infraorbital foramen to supply the surrounding area of face. This branch also supplies the incisor and canine teeth and gums in addition to the maxillary sinus. Anastomoses between these branches and branches of the facial (external carotid supply) and ophthalmic (internal carotid supply) arteries are plentiful. On a prosection of the side wall of the nose locate the palatine arteries which pass through the palatine canal to supply the palate and roof of the mouth. A branch passes backward to supply the nasopharynx and the **sphenopalatine artery** supplies the side walls and septum of the nose.

Arterial anastomoses of the head and neck (6.10.7)

Anastomoses are profuse in the head, neck, and, in particular, the scalp and face; they exist between the right and left sides, between the arteries inside and outside the skull, and between the deep and superficial supplies of the neck and face. When the scalp is lacerated, bleeding is profuse not only because it is well supplied with blood vessels but also because of the fibrous nature of the subcutaneous tissue of the scalp which tends to prevent rapid contraction of the severed arterioles. Nevertheless, a scalping injury can usually be repaired provided that a single vascular pedicle is left.

The function of the circle of Willis (**6.10.4**), which is an anastomosis between the internal carotid and vertebral arteries at the base of the brain, is to equalize the blood pressure on either side of the brain. However, the circle is sometimes incomplete or inadequate.

= = = = = from internal carotid

supratrochlear, supraorbital, and superficial temporal

superficial and deep arteries of face

septal anastomosis

labial arteries of both sides

inferior alveolar, sublingual, and submental

superior and inferior thyroid

deep temporal and middle meningeal

superficial temporal arteries of both sides

superficial temporal, posterior auricular, and occipital

occipital and deep cervical

ascending cervical, vertebral, and deep cervical

6.10.7
Principal anastomoses between vertebral, internal, and external carotid arteries.

Qu. 10H *It is sometimes possible to ligate the common carotid artery without evidence of ensuing brain damage. What anastomoses would allow continuing blood flow in the internal carotid artery?*

Venous drainage

Intracranial and cranial veins

Venous drainage of the brain, eye, and interior of the skull is conducted through the intracranial venous sinuses within the dura mater. These have neither smooth muscle nor valves in their walls, and are held open by the attachment of dura to bone. In general, veins of the head and neck lack valves, and rely on gravity and the negative intrathoracic pressure developed at inspiration for the return of blood to the heart.

Qu. 10I *What is the pressure of blood in the intracranial venous sinuses?*

Cerebrospinal fluid, secreted by the choroid plexuses into the ventricles of the brain, leaves the ventricular system, circulates through the subarachnoid space, and finally drains through arachnoid granulations into the superior sagittal sinus. The circulation of CSF is passive since the hydrostatic pressure is higher and the osmotic pressure lower in the CSF than in the venous blood of the sinus.

Qu. 10J *What mechanisms make the hydrostatic and osmotic pressures of the CSF different from those in the venous sinus?*

The arrangement of the various intracranial venous sinuses has already been considered (p. 79). They communicate with veins of the scalp through **emissary veins**. Examine the posterior aspect of the mastoid process, and the posterior aspect of the foramen magnum and locate the mastoid emissary and posterior condylar foramina from which certain of the emissary veins emerge. The cavernous sinuses, which drain part of the lateral aspect of the brain, the pituitary, and also the ophthalmic vein from the orbit, communicate with veins of the face and the pterygoid plexus of veins in the infratemporal fossa. However, the blood in the cranial venous sinuses drains out of the cranium principally via the sigmoid sinus which emerges from the jugular foramen to form, with the inferior petrosal sinus, the **internal jugular vein**. Examine the jugular foramen of a skull and note the rounded impression for the expanded **bulb** of the internal jugular vein and its proximity to the middle ear (see **6.10.14**).

Bone marrow fills the potential space (diploë) between the outer and inner layers of compact bone of the cranial bones. Draining the diploë are **diploic veins** which communicate with veins of the scalp, intracranial venous sinuses, and meningeal veins. The meningeal veins drain into the venous sinuses, diploic veins and also, through the base of the skull, into the pterygoid plexus. There is thus considerable communication between the intra- and extracranial veins.

Extracranial veins of the head and face (6.10.8)

Identify the various superficial veins of the head and face shown in **6.10.8**, remembering that variation of venous drainage is common. Locate the **supratrochlear** and **supraorbital veins** which drain the forehead; they empty into the facial vein but also communicate with ophthalmic veins draining the orbit into the cavernous sinus. The **facial vein** is prominent and runs with the facial artery from the medial angle of the eye to the inferior border of the mandible, draining most parts of the face including its deep aspects. The remainder of the scalp is drained by **superficial temporal**, **posterior auricular**, and **occipital** veins. The occipital vein pierces the investing layer of cervical fascia and trapezius to join the vertebral vein. The superficial temporal vein passes down in front of the ear, receives veins which drain the side of the face, and is joined by the **maxillary veins** which drain the **pterygoid plexus** in the infratemporal fossa, to form the **retromandibular vein** in front of the tragus. This usually divides into two: its anterior limb joins the facial vein just before it pierces the investing layer of cervical fascia to enter the **internal jugular vein**; its posterior limb usually joins the posterior auricular to form the **external jugular vein**.

The major veins of the neck lie either superficial or deep to the investing layer of deep fascia. Remember that, in the face, there is no such separation because the facial muscles attach to the skin. Superficial to the investing layer of cervical fascia are two vertically running veins. Locate the **external jugular vein**; as it descends it first crosses sternomastoid and then, just above the clavicle, pierces the investing layer of cervical fascia to drain into the **subclavian vein**. The **anterior jugular vein** drains superficial aspects of the chin and neck including the larynx; it runs vertically downward towards the manubrium, pierces the deep fascia and forms an arch with the vein opposite, and then turns laterally to join the external jugular as it enters the subclavian vein. The **internal jugular vein** lies deep to the investing layer of fascia in the neck. It begins in the base of the skull as a continuation of the sigmoid sinus and runs vertically downward in the neck, enclosed with the internal and common carotid artery and the vagus in the carotid sheath. It joins the subclavian vein behind the manubriosternal joint to form the brachiocephalic vein.

Qu. 10K *For what reason is the carotid sheath thinner over the vein than over the artery?*

The internal jugular veins receive tributaries which correspond to most branches of the external carotid artery. The **vertebral vein** accompa-

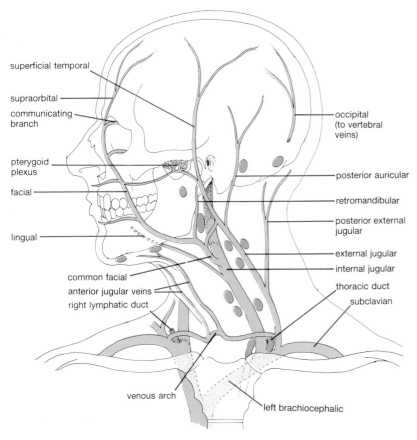

6.10.8
Venous drainage of head and neck (intracranial sinuses not shown); position of major groups of lymph nodes.

nies the vertebral artery in most of its course; it drains the posterior aspect of the scalp and drains the muscles, vertebrae, and spinal cord of the cervical region, finally emerging through the vertebral triangle to empty into the brachiocephalic vein. It does *not* drain the parts of the brain supplied by the vertebral artery.

Lymphatic drainage (6.10.9)

Much of the head and neck has a profuse lymphatic drainage as it is associated with the skin, upper alimentary tract, and respiratory tract. However, neither the central nervous system nor the cornea of the eye have lymphatics. This means that the cornea can be transplanted without problems of immunological rejection.

The lymphatic drainage of the head and neck follows the general pattern, in that **superficial** lymphatic vessels run with the superficial veins within the superficial fascia (i.e. superficial to the investing layer of deep cervical fascia; see **6.2.12**). The lymphatics drain into clusters of glands which lie in a 'collar' arrangement around the neck at the upper attachment of the investing layer of deep cervical fascia.

Locate lymph nodes of:

- The **submental group** which lie below the symphysis menti and drain the front of the scalp, the teeth, and the anterior part of the tongue and face.

- The **submandibular group** which lie around the submandibular salivary gland and drain the side of the scalp and face, and receive from the submental nodes.

- The **parotid group** which lie superficial to the parotid gland and drain the side of the scalp and the face.

- The **mastoid group** behind the ear and the **occipital group** at the base of the occiput which both drain the adjacent scalp.

These nodes either drain through the deep fascia into **upper deep cervical nodes** around the internal jugular vein, or along vertically running superficial lymph vessels with associated nodes which lie along the external jugular and anterior jugular veins, and drain eventually into **lower deep cervical nodes** at the root of the neck. The deep cervical group lie in a vertical chain around the internal jugular vein but tend to group behind the angle of the mandible and at the root of the neck.

The **deep** lymphatic drainage from the orbit, nose, nasopharynx, and oropharynx passes with blood vessels to a ring of lymphatic tissue which lies around the oropharynx.

- The **lingual tonsil** lies at the base of the tongue.

- The **palatine tonsil** lies between the pillars of the fauces.

- **Retropharyngeal lymphatic tissue** lies deep to the pharyngeal mucous membrane. In young children lymphatic tissue is prominent around the pharyngeal opening of the pharyngotympanic tube and in the nasopharynx (**adenoids**) where it can become inflamed. The lymphatic tissue which protects the entrance passages to the larynx and pharynx tends to regress by about 7 years of age.

Qu. 10L *If the adenoids become enlarged as a result of inflammation, what effect does this have on respiration?*

This deep lymphatic tissue (and the superficial nodes) of the head and neck drains into deep cervical glands which drain either into the **right cervical trunk** which empties into the junction between the right internal jugular and subclavian veins or into the **left cervical trunk** which enters the thoracic duct before emptying into the venous junction on the left side.

The lymphatic drainage of the **tongue** is important because the tongue is often involved in maligant or non-malignant ulceration. It drains to submental, submandibular, and upper deep cervical nodes (p. 59; extensive anastomoses allow lymph to reach any of these nodes on either side).

Qu. 10M *Where would you expect to find infection if glands were enlarged and painful in: (a) the occipital region; (b) the submental group?*

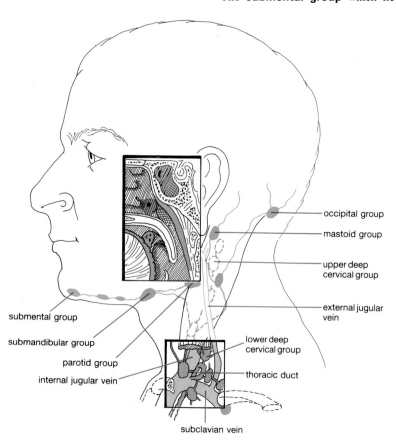

occipital group

mastoid group

upper deep cervical group

external jugular vein

lower deep cervical group

thoracic duct

subclavian vein

submental group

submandibular group

parotid group

internal jugular vein

6.10.9
Lymph nodes, lymphatic tissue, and main lymphatic trunks of head and neck (view of pharynx enlarged).

C. Radiology

Examine both the lateral and A–P views of a vertebral angiogram (**6.10.10**). Trace the course of the artery vertically upward through the foramina transversaria to the axis. Note that it passes laterally to enter the foramen in the transverse process of the atlas and then medially across the lateral mass of the atlas before entering the foramen magnum. The right and left vertebral arteries (RV, LV) then join within the skull to form the basilar artery (BA). Identify the posterior cerebral vessels (PC) which are terminal branches of the basilar; note also the branches in the posterior cranial fossa which supply the cerebellum. The posterior communicating (P) branches of the circle of Willis are

visible, but are seen more easily on the A–P view. Note also the difference in size of the right and left vertebral arteries (see **6.15.1d**).

Examine the lateral view of an internal carotid angiogram (**6.10.11a**). Locate the region of the carotid canal and trace the course of the artery through the petrous temporal bone to the foramen lacerum and on to the sides of the sphenoid. Identify the ophthalmic artery, and the middle (MC) and anterior cerebral (AC) arteries. Now turn to the P–A view (**6.10.11b**) which is unlabelled, and identify the arteries you have seen in **6.10.11a**. Note that contrast medium has not reached the contralateral side of the brain. **6.10.11c** is from the same subject but is an A–P view of the venous phase. Note the cerebral veins (CV) running anteriorly to

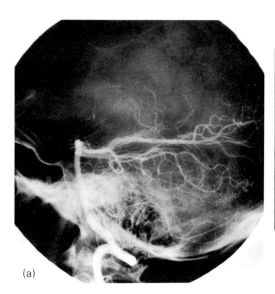
(a)

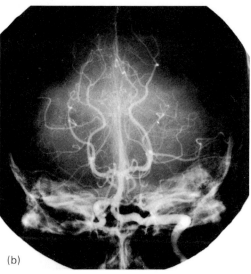
(b)

6.10.10
Angiograms of vertebral artery: (a) lateral view; (b) A–P view.

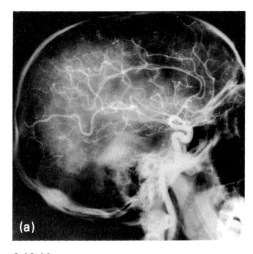
(a)

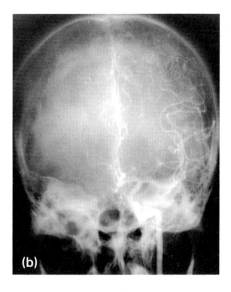
(b)

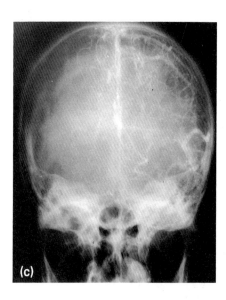
(c)

6.10.11
Angiograms of internal carotid artery: (a) lateral view: (b) P–A view; (c) venous phase, A–P view.

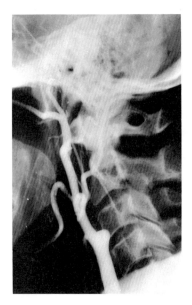

6.10.12
Carotid angiogram (abnormal).

enter the superior sagittal sinus; an inferior cerebral vein (IC) entering the straight sinus.

Examine **6.10.12** and identify the branches of the external carotid artery.

Qu. 10N *What is abnormal about radiograph* **6.10.12**?

Examine angiogram **6.10.13** in which contrast material has been introduced into the internal carotid artery.

Qu. 10O *What unexpected features do you see in* **6.10.13**?

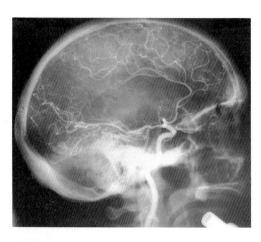

6.10.13
Internal carotid angiogram (abnormal).

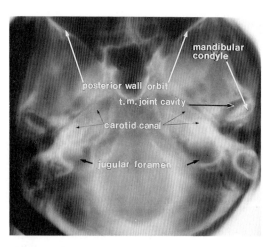

6.10.14
Planar tomogram of skull base showing carotid canal and jugular foramen.

Planar basal tomograms of the skull (**6.10.14**) can be used to visualize the carotid canal (C) and jugular foramen (J).

Examine **6.10.15**, a coronal CT of the region where the internal carotid artery emerges from the cavernous sinus above the sphenoid bone and its (divided) air sinus. Contrast medium has been introduced into the CSF to outline vessels and other structures in the subarachnoid space. Note the internal carotid artery (I) bifurcating into the middle and anterior cerebral arteries, the optic chiasm and floor of hypothalamus (C), the pituitary stalk (S), and position of the pituitary gland (P) between the cavernous sinuses.

Requirements:

Articulated skeleton, separate skull, and skull cap.

Prosections of the subclavian and vertebral vessels; common and internal carotid arteries; external carotid and its branches; circle of Willis; venous sinuses of the cranial cavity; venous drainage of the face and scalp; external veins of the face and neck; deep venous drainage (internal jugular vein) of the head and neck; sagittal section of the head and neck to show position of the pharyngeal and lingual tonsillar tissue and lymphatic tissue of the nasopharynx.

Arteriograms of the carotid; subclavian and vertebral trunks.

Diagrams of the lymphatic drainage of the head and neck.

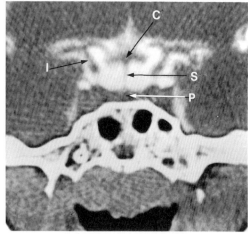

6.10.15
Coronal CT with intrathecal contrast showing bifurcation of internal carotid artery within subarachnoid space

Seminars 11 – 14

Innervation of the head and neck

Introduction

The nerve supply to the various regions of the head and neck has already been considered. Seminars 11–14 focus on the twelve paired cranial nerves and eight paired cervical nerves originating from the cranial and cervical parts of the central nervous system. The first two cranial nerves (olfactory and optic) are more accurately thought of as extensions of the brain. The cranial and cervical nerves contain a variety of fibre types: sensory fibres to skin, connective tissues, and derivatives of the foregut; motor and sensory fibres to striated muscle; and preganglionic parasympathetic fibres for the supply of salivary, lacrimal, and mucous glands and some smooth muscle. By contrast, the sympathetic supply to glands and smooth muscle originates outside the head and neck, from the thoracic spinal cord. The following *summary* of the organization of the innervation of the head and neck should be re-read after seminar 14 has been completed.

Sensory innervation

Somatic sensory (afferent) neurons inner-vate skin, muscles, joints, and connective tissues from which information (touch, proprioception, distension, pain, temperature) can be trans-duced. Their cell bodies are located in the sen-sory ganglia of cranial nerves (V, VII, IX, X) and in the dorsal root ganglia of cervical nerves; their central processes pass into the brainstem or cervical spinal cord to synapse in the appropri-ate sensory nuclei. These nuclei develop in the lateral (alar lamina) part of the neural folds; closure of the spinal neural tube carries the cer-vical cells to a dorsal position, but in the brain-stem most remain laterally situated. From these nuclei some 'second-order' neurons project to the thalamus, whence information is carried by 'third-order' neurons to the sensory cortex of the cerebrum and other areas. At each synaptic relay the information undergoes significant pro-cessing.

Visceral sensory neurons innervate mechano-, noci-, and chemoreceptors of visceral organs. Their cell bodies are located in ganglia (equiva-lent to dorsal root ganglia) of certain cranial nerves (VII, IX, X) where the nerves pass through the skull. The central processes of the ganglion cells synapse in a nucleus in the medul-la (nucleus solitarius), from which second-order neurons project information to higher centres.

Motor innervation to striated muscle

In the developing central nervous system **motor neurons** form a continuous column of cells in the ventral aspect (basal lamina) of the spinal cord (the putative ventral horn) and clusters of cells in the medial aspect of the brainstem. Their axons collect to form the motor component of cranial nerves which supply the various groups of striated muscles of the head and neck. Cra-nial nerves, III, IV, and VI supply the extrinsic ocular muscles; V, VII, IX, X, and XI supply muscles of branchial arch origin; XII supplies muscles of the tongue; and ventral roots of cer-vical nerves supply somite-derived muscles of the neck. Muscles of the pharynx, larynx, and palate are all supplied from a large group of neurons, the nucleus ambiguus, which is situated in the medulla of the brainstem between the sensory nuclei and the more medially placed motor nuclei.

Autonomic innervation

Parasympathetic innervation. Preganglionic neurons of the parasympathetic system form nuclei in the brainstem (and also in the sacral spinal cord). Their axons leave the central ner-vous system with the oculomotor (III), facial (VII), glossopharyngeal (IX), and vagus (X) nerves (and with sacral ventral roots) and pass peripherally to synapse on ganglion cells which are collected into **cranial parasympathetic ganglia** (ciliary — III, pterygopalatine — VII, submandibular — VII, otic — IX) or found in small groups in the walls of viscera (X). From the ganglia, postganglionic parasympathetic fibres pass to their targets. Cells of the **ciliary ganglion** supply sphincter pupillae and ciliaris within the eye; the **pterygopalatine ganglion** supplies the lacrimal gland and mucus-/serous-secreting cells in the nose, nasopharynx, and palate; the **submandibular ganglion** supplies the submandibular and sublingual salivary glands and scattered secretory cells in the floor of the mouth; the **otic ganglion** supplies the parotid gland and the vestibule of the mouth; mucus- and serous-secreting cells in the pharynx, oesophagus, larynx, and trachea are supplied by the glossopharyngeal and vagus nerves. The vagus nerve also supplies pregang-lionic parasympathetic fibres to the heart and lungs in the thorax, and to the alimentary tract and its derivatives as far as the splenic flexure of the colon in the abdomen.

Sensory and postganglionic sympathetic fibres pass through, but do not synapse in, the cranial parasympathetic ganglia and are distributed with their branches.

Sympathetic system. Preganglionic sympathetic neurons form a column of cells (lateral horn) in the thoracic spinal cord (T1–L2). Therefore, the sympathetic innervation to the entire head and neck originates from the thorax (T1–T4). The preganglionic sympathetic fibres leave the spinal cord with T1–L2 nerve roots and project (via white rami communicantes) to ganglion cells of the sympathetic chain from which postganglionic sympathetic fibres pass to their targets. Within the head and neck the sympathetic chain forms three large ganglia, the superior, middle (often very small), and inferior cervical ganglia. The inferior ganglion is often fused with the T1 ganglion and called the 'stellate' ganglion. The sympathetic postganglionic fibres in the head and neck innervate smooth muscle of blood vessels, dilator pupillae, and part of levator palpebrae, salivary glands, and serous-/mucus-secreting cells of the head and neck. They travel to their targets largely along blood vessels, but are also distributed as grey rami communicantes through the cervical spinal nerves or via the cranial parasympathetic ganglia.

The motor, sensory, and autonomic nuclei within the central nervous system all receive numerous inputs from higher centres of the central nervous system.

The sagittal section of the head (**6.11.1**) shows the position of the brainstem in relation to the skull and face. Note the origin of all the cranial nerves.

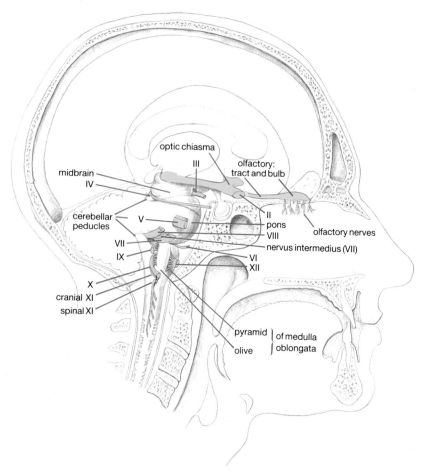

6.11.1
Origin of cranial nerves from brainstem.

Seminar 11

Cranial nerves: olfactory; optic; oculomotor, trochlear and abducent; ophthalmic division of trigeminal

The **aim** of this seminar is to study the function, origin, course, and distribution of the olfactory nerve (I), and the cranial nerves within the orbit, viz. the optic (II), oculomotor (III), trochlear (IV), ophthalmic division of the trigeminal (V), and the abducent (VI) cranial nerves.

A. Living anatomy

Olfactory nerve (I)

In most non-primate mammals the sense of smell is of paramount importance, both for self-preservation with regard to feeding and signalling of predators, but also for the propagation of the species. Olfactory signals (pheromones) and the sense of smell, which is of demonstrable importance in monkeys, may well play a larger part in human behaviour than we give it credit for. The sense of taste is closely associated with that of smell and very often diminishes as people grow older.

The olfactory nerve can be tested by asking the subject to identify various common smells, such as coffee, impregnated into cotton wool.

Optic nerve (II)

Man is a vision-dominated animal, and a large amount of the central nervous system is concerned with the analysis of visual input. Consider mechanisms which underlie visual acuity, colour vision, intensity, and depth perception.

To test the field of vision, sit opposite your partner and get him to fix his gaze on your nose; then cover his left eye with your right hand. Starting at arm's length, gradually bring the moving fingers of your left hand into your partner's field of vision at different points on the radius of a circle. Make a plot of the field of vision and then repeat the exercise covering the other eye, noting whether the fields of vision are similar on both sides.

Qu. 11A *Why is the field of vision less circular on the medial side?*

If you do the same with a small coloured object, you will find that your partner can perceive the presence of the object before its colour can be identified. This is because the colour-sensing cones are concentrated in the centre of the field of vision, the fovea. Visual acuity is tested with charts of letters of gradually diminishing size. To test the pupillary light reflexes and accommodation reflexes, see p. 86.

Oculomotor (III), trochlear (IV) and abducent (VI) nerves

These nerves supply the muscles that move the eye. Ask your partner to keep his head still but follow the movements of your finger to the upper, lower, right, and left periphery of the visual field. The abducent (VI) nerve is tested by asking the subject to look laterally; the oculomotor (III) by looking medially and checking that full retraction of the upper lid occurs when the subject looks up; the trochlear (IV) nerve is tested by the subject attempting to look downward when already gazing laterally.

Qu. 11B *Explain these tests on the basis of the muscles innervated by each nerve.*

Ophthalmic division of the trigeminal nerve (V)

To test the ophthalmic division of the trigeminal nerve, gently touch the skin of the forehead of your partner with a wisp of cotton wool and then with a pin and establish that touch and a pricking sensation are elicited. Also, touch the cornea very gently with cotton wool to elicit a blink (corneal) reflex.

B. Prosections

Olfactory nerve (I) (6.11.1, see 6.3.7)

The olfactory nerve arises from specialized neuroepithelial cells (**olfactory epithelium**) which are found in the roof and superior conchae of the nose. The olfactory neuroepithelial cells can be considered as bipolar neurons although, unlike neurons, they are a self-replacing population. Lipid-soluble (and some non-lipid-soluble) molecules in the inhaled air impinge on ciliated receptors covering the luminal surface of the cells and stimulate the sense of smell. The unmyelinated axons pass through the cribriform plate of the ethmoid bone to synapse in the olfactory bulb.

Qu. 11C *In which cranial fossa does the olfactory bulb lie?*

The output (mitral) cells of the olfactory bulb have axons which project along the olfactory tract to areas of the temporal and frontal cerebral cortex where olfactory information is integrated. If the cribriform plate is fractured and

the sense of smell lost, food can taste uninteresting and even unpalatable.

Optic nerve (II) (6.11.1, 6.11.2, see 6.8.17)

Axons from retinal ganglion cells cross the surface of the retina to reach the **optic disc** where they are collected together as the **optic nerve** (see **6.8.5**). Examine a prosection of the orbit and note that the optic nerve, lying within its sheath of dura and arachnoid, passes through the optic foramen (with the ophthalmic artery) to reach the anterior cranial fossa where the two nerves join to form the **optic chiasm**. In the chiasm, axons from the medial (nasal) aspect of the retina cross the midline to join axons from the lateral (temporal) aspect of the retina of the opposite eye, thereby segregating information from the left and right **visual fields** into, respectively, the right and left **optic tracts**.

Qu. 11D *If when testing the visual fields you detected a large defect in both temporal fields of vision, what might be the cause?*

Follow the tracts as they pass posteriorly around the upper part of the brain stem. A few fibres end in the midbrain (pretectal nucleus and superior colliculus) for visual reflexes, but the vast majority, which serve visual perception, end in the lateral geniculate nuclei of the thalamus. The further course of the visual pathway can be seen on a horizontal section through the brain (**6.11.2**). Axons of lateral geniculate neurons sweep posteriorly around the lateral ventricle of the brain as the **optic radiation** to reach the **visual cortex** of the occipital pole of the brain.

Qu. 11E *What would be the result of damage to (a) the right optic nerve? (b) the right optic tract?*

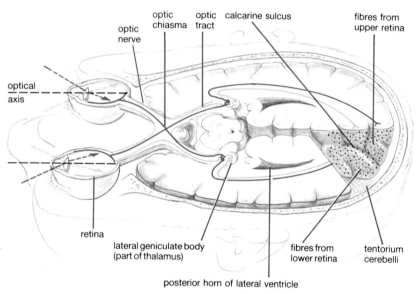

6.11.2
Visual pathways.

Oculomotor nerve (III) (6.11.1, 6.11.3; see 6.11.8 and 6.8.17)

This nerve arises from motor neurons in the oculomotor nucleus of the midbrain. Follow the nerve of each side as it emerges from the medial aspect of the cerebral peduncle of the brainstem and runs forward in the subarachnoid space (interpeduncular fossa) to pierce the roof of the cavernous sinus (see **6.7.7**). The nerve traverses the lateral wall of the sinus and passes through the superior orbital fissure within the tendinous ring. Within the orbit it divides into a superior branch which passes upward through superior rectus to reach levator palpebrae, and an inferior branch to the medial and inferior recti and to inferior oblique.

The branch to inferior oblique carries **parasympathetic fibres** originating in a part of the oculomotor nucleus (Edinger–Westphal nucleus). These fibres pass to the **ciliary ganglion** where they synapse, and from which postganglionic axons run in the short ciliary nerves to enter the eyeball and supply the ciliary muscle and sphincter pupillae.

Examine photographs **6.11.4a,b** of a subject with paralysis of the III nerve on the right side.

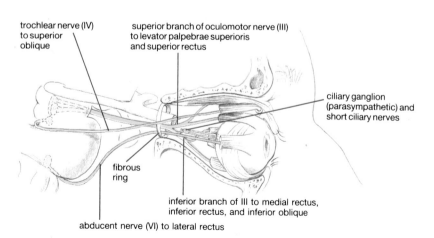

6.11.3
Oculomotor, abducent, and trochlear nerves.

In (b) the eyelid has been raised to show the divergent position of the two eyes and the dilated pupil.

Qu. 11F *Explain (a) the ptosis, (b) the divergence, and (c) the dilated pupil.*

Trochlear nerve (IV) (6.11.1, 6.11.3, see 6.7.7)

The trochlear nerve supplies a single extraocular muscle, superior oblique. The nerve is very slender and is the only one to emerge from the dorsal aspect of the brainstem. Locate the nerve as it emerges from the dorsal aspect of the midbrain, and passes laterally around the brainstem within the subarachnoid space to pierce the dura of the cavernous sinus immediately posterior to the oculomotor nerve. It passes forward in the lateral wall of the cavernous sinus and through the lateral aspect of the superior orbital fissure to enter the orbit where it runs outside the muscular sheath of the eyeball and supplies superior oblique.

Qu. 11G *What would be the effect of damage to the trochlear nerve?*

Abducent nerve (VI) (6,11.1, 6.11.3, see 6.7.7)

The abducent nerve also supplies a single extraocular muscle, lateral rectus. Its cell bodies are situated in the brainstem at the level of the pons, at a position analogous to the nuclei of III and IV. Locate the abducent nerve as it emerges from the brainstem close to the midline at the junction of the pons and medulla; it runs through the subarachnoid space to pierce the dura over the basisphenoid and run up over the apex of the petrous temporal bone and into the cavernous sinus in which it lies lateral to the internal carotid artery. It then passes through the superior orbital fissure to supply the lateral rectus from within the muscular sheath of the eyeball.

Qu. 11H *What would be the effect of damage to the abducent nerve (6.11.5)?*

Ophthalmic division of the trigeminal nerve (V) (6.11.6, 6.11.7)

The ophthalmic division of the trigeminal nerve contains only sensory fibres, the cell bodies of which are situated in the trigeminal ganglion (equivalent to a dorsal root ganglion) which lies in a recess of dura mater on the apex of the petrous temporal bone (this excludes proprioceptive fibres, cell bodies of which lie in the midbrain).

The three major divisions of the trigeminal nerve arise from the ganglion. Trace the ophthalmic division as it runs along the lateral wall of the cavernous sinus inferior to the oculomotor and trochlear nerves, and divides into frontal, lacrimal, and nasociliary branches which enter the orbit through the superior orbital fissure. Locate the large **frontal nerve** as it runs

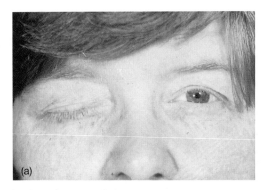

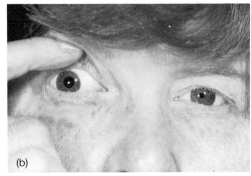

6.11.4
Right oculomotor nerve palsy. In (b) the eyelid has been raised to show divergent eyes and dilated pupil.

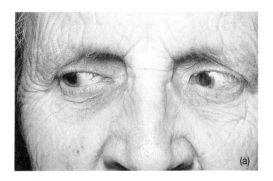

6.11.5
Left abducent nerve palsy. Patient looking to her (a) right and (b) left; the left eye fails to abduct.

forward beneath the roof of the orbit on the dorsal surface of levator palpebrae, where it divides into **supratrochlear** and **supraorbital** branches. The supraorbital is the larger and

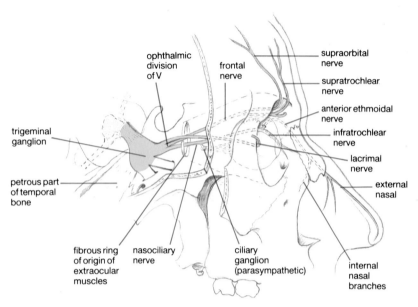

6.11.6
Ophthalmic division of trigeminal nerve (V).

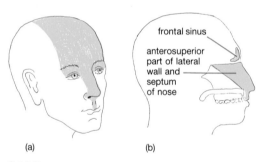

(a)   (b)

6.11.7
Sensory distribution of ophthalmic division of V:
(a) to face and scalp; (b) to lacrimal gland, nose, and frontal sinus.

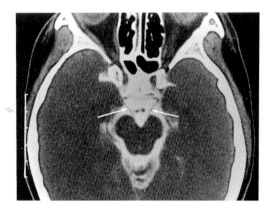

6.11.8
Horizontal CT showing oculomotor nerves
(arrows) in subarachnoid space.

more lateral and passes through the supraorbital fissure (or foramen) on to the forehead which it supplies. Locate the **lacrimal nerve** which passes laterally outside the muscular sheath of the eyeball to supply sensory fibres to the lacrimal gland, conjunctiva, upper eyelid, and lateral forehead. *En route*, it is joined by postganglionic parasympathetic fibres from the pterygopalatine ganglion (VII nerve) which, with sympathetic fibres derived from the carotid plexus, regulate the secretion of lacrimal fluid.

Identify the **nasociliary nerve** as it crosses the optic nerve to reach the medial part of the orbit (see **6.8.17**). It gives sensory **ciliary** branches to the eyeball (some of which run through the ciliary ganglion though they do not synapse there) and supplies the ethmoid and sphenoidal air sinuses, and ends by dividing into an **infratrochlear** branch which leaves the medial wall of the orbit to supply the skin of the root of the nose, and the **anterior ethmoidal nerve**. The latter passes through the anterior ethmoidal foramen, supplies ethmoidal air cells, emerges on to the cribriform plate, and then passes downward at the side of the crista galli to enter the nose. It supplies the upper anterior part of the septum and lateral wall (**internal nasal** branches) of the nose and finally emerges between the nasal bone and cartilage as the **external nasal nerve** to supply the tip of the nose.

Within the orbit, the ciliary nerves usually carry postganglionic sympathetic fibres (from the internal carotid plexus) to the ciliary ganglion and eyeball, where they supply dilator pupillae and blood vessels.

In summary, the ophthalmic division of the trigeminal nerve supplies sensory fibres to most of the skin of the forehead, upper eyelids, bridge of the nose, and philtrum of the upper lip; and structures of the face lying deep to this area, i.e. the lacrimal glands, conjunctiva, frontal sinuses, and the anterosuperior part of the nasal cavity.

Qu. 11I *Sensory neurons can become invaded by the herpes zoster virus, with painful blistering in the sensory distribution of the nerve. Where would the blisters be located if the ophthalmic division of the trigeminal nerve were to be involved?*

C. Radiology

Examine the CT **6.11.8** which has been taken at an angle to the horizontal to show structures in the area around the base of the brain and pituitary stalk. Note the oculomotor nerve (arrowed) on either side emerging from the medial aspect of the cerebral peduncles.

Requirements:
    Skull with removable skull cap.
    Prosections of a sagittal section of the head showing the olfactory bulb and tract; III, IV, and the internal carotid artery in relation to the cavernous sinus; the orbit (used in Seminar 8).

Seminar 12

Cranial nerves: maxillary and mandibular division of trigeminal; facial; vestibulocochlear

The **aim** of this seminar is to study the function, origins, course, and distribution of the maxillary and mandibular divisions of the trigeminal nerve (V) and the facial nerve (VII), which supply the nose, mouth, cheeks, and muscles of the face, and also the vestibulocochlear nerve (VIII) which supplies the special sense organs of the inner ear.

A. Living anatomy

Maxillary and mandibular divisions of the trigeminal nerve (V)

To test the **sensory** distribution, establish touch and pain sensitivity over the lower eyelid, upper cheek, and upper lip (maxillary division) and over the the lower cheek and chin (mandibular division) (**6.12.1**). To test the **motor** distribution (which is confined to the mandibular division) ask the subject to clench his jaws, and feel the contraction of masseter.

Facial nerve (VII)

Ask your partner to 'screw up' his eyes, and to show his teeth to test the **motor** component of the facial nerve (both are tested because facial nerve neurons in the brainstem supplying orbicularis oculi are supplied from both left and right cerebral cortices, whereas those to the orbicularis oris are supplied from only the contralateral cortex). To check the **secretomotor** component, establish that the conjunctiva are moist with lacrimal secretions and that saliva emerges from both submandibular ducts in the floor of the mouth.

Vestibulocochlear nerve (VIII)

To test the auditory system in a simple way, place a vibrating tuning fork beside each ear in turn. Hearing the vibrations relies on conduction of sound through air and the ossicle chain in addition to the function of the internal ear and the cochlear component of VIII. Now place the base of a vibrating tuning fork against each mastoid process so that the sound is conducted through bone to the inner ear.

Qu. 12A *If the sound cannot be heard when conducted through air but can be heard when conducted through bone, what would you conclude?*

To compare the auditory ability of the two inner ears, place a vibrating tuning fork on the centre of the forehead. The sound should be perceived as equally loud on both sides.

Exposure to prolonged loud noise can result in auditory loss at some frequencies but not others. To test this, the auditory threshold for sounds of different frequencies must be established.

The role of the vestibular apparatus in generating sensations related to equilibration can be demonstrated by pouring water at a little above body temperature into one external auditory meatus. Thermal convection currents set up in the fluid within the vestibular apparatus give a sensation of disordered balance.

B. Prosections of the trigeminal and facial and vestibulocochlear nerves

Maxillary division of the trigeminal nerve (6.11.1, 6.12.2)

The maxillary nerve is entirely sensory in function, like the ophthalmic nerve. On a prosection of the middle cranial fossa trace the maxillary nerve as it passes from the trigeminal ganglion and through the foramen rotundum. Within the skull it supplies the surrounding dura mater.

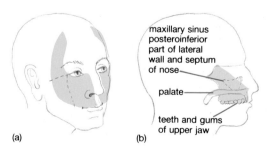

(a)   (b)

6.12.1
Sensory distribution of maxillary division of V:
(a) to face and scalp; (b) to lacrimal gland, nose,

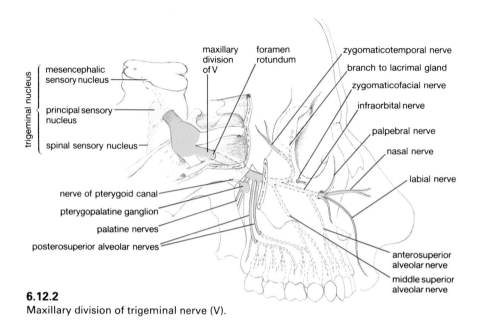

6.12.2
Maxillary division of trigeminal nerve (V).

Pass a probe through the foramen rotundum of a skull and note that it reaches the pterygomaxillary fissure. The maxillary nerve runs through the upper part of the fissure where it divides into a number of branches, and continues through the inferior orbital fissure on to the floor of the orbit where it sinks into the infraorbital canal as the **infraorbital nerve**.

Examine prosections of the upper jaw and find the **posterior superior alveolar nerve** which passes downward over the posterior aspect of the maxilla, which it pierces to supply the upper molar teeth and adjacent gum and cheek. A **zygomatic nerve** arises within the fissure, enters the orbit, and passes on to its lateral wall where it divides into zygomaticofacial and zygomaticotemporal branches which supply skin of the cheek over the zygoma in the temporal region. In the floor of the orbit, **middle** and **anterior superior alveolar nerves** arise from the infraorbital nerve. These traverse and supply the lateral and anterior walls of the maxillary sinus to reach the premolar and incisor teeth and associated gum which they supply.

Qu. 12B *If the maxillary sinus is infected (see 6.3.20) where might pain be felt?*

The infraorbital nerve finally emerges through the infraorbital foramen to supply the lower eyelid and conjunctiva, the cheek and upper lip.

On a skull, probe the pterygomaxillary fissure and demonstrate its continuity with a recess, the **pterygopalatine (sphenopalatine) fossa**, in the side wall of the nose immediately posterior to the middle concha. Within the fossa lies the **pterygopalatine ganglion** which hangs below and receives sensory fibres from the maxillary nerve. The preganglionic parasympathetic fibres to the ganglion are derived from the facial nerve via its greater superficial petrosal branch. This joins with postganglionic sympathetic fibres of the internal carotid plexus (deep petrosal nerve)

in the foramen lacerum, and passes forward through the pterygoid canal in the sphenoid bone to reach the pterygopalatine ganglion. Like the branches of the ciliary ganglion, those of the pterygopalatine ganglion distribute sensory, parasympathetic, and sympathetic fibres, but only the parasympathetic fibres synapse in the ganglion. On a prosection of the nose, locate the **greater** and **lesser palatine nerves** which descend through canals in the palatine bone to supply the hard and soft palate. **Posterior nasal nerves** supply the side wall and septum of the nose, and a **pharyngeal branch** passes backward to supply the roof of the nasopharynx. Trace the **nasopalatine nerve** as it arches over the roof of the nose to pass diagonally forward and downward on the nasal septum, which it supplies, and enters the incisive foramen behind the incisor teeth to supply the anterior aspect of the hard palate. The pterygopalatine ganglion also supplies secretomotor fibres to the lacrimal gland. These postganglionic parasympathetic fibres join the zygomatic branch of the maxillary nerve but leave it in the orbit to join the lacrimal nerve for distribution to the lacrimal gland.

Qu. 12C *What would result if this innervation were damaged? And where could this occur?*

In summary, the maxillary division of the trigeminal nerve supplies most of the skin of the cheek and upper lip and all structures of the face lying deep to this area (i.e. the oral aspect of the hard and soft palate, teeth and gums of the upper jaw, the maxillary sinus, and the posteroinferior part of the nasal cavity).

Mandibular division of the trigeminal nerve (V) (6.12.3–6.12.5)

The mandibular division of the V nerve is a mixed nerve. Its large sensory root leaves the trigeminal ganglion and is joined by a motor

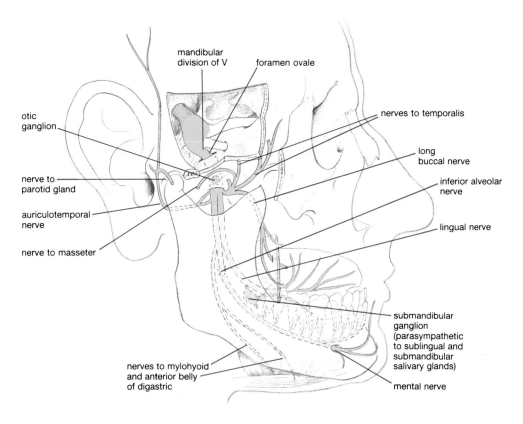

6.12.3
Mandibular division of trigeminal nerve (V).

root, the cell bodies of which lie in the pontine part of the brainstem. The mandibular nerve leaves the cranial cavity through the foramen ovale to reach the infratemporal fossa in which it divides almost immediately into a series of branches, including three main sensory branches, the auriculotemporal, lingual, and inferior alveolar nerves.

On a skull, pass a probe through the foramen ovale and remind yourself of the surrounding bony features inside and outside the skull. As the mandibular nerve leaves the skull it gives a small meningeal branch, the nervus spinosus, which accompanies the middle meningeal artery through the foramen spinosum to supply sensation to the dura mater of the middle cranial fossa and the mastoid air cells.

On a prosection of the infratemporal fossa, identify the **auriculotemporal nerve** and follow it laterally and upward behind the temporomandibular joint to where it supplies sensory fibres to the skin over the temple and the parotid gland. It supplies the tragus and external aspect of the pinna above; it also supplies the upper, anterior part of the external auditory meatus and tympanic membrane. The branch to the parotid gland also carries secretomotor fibres from the otic ganglion to the parotid gland (see below). Locate the **buccal nerve** which supplies the skin and mucous membrane of the cheek.

Trace the **lingual nerve** as it passes downward over medial pterygoid and then forward in the gap between the superior and middle con-

strictors of the pharynx, to lie between mylohyoid and hyoglossus, and reach the mucosa of the tongue.

Qu. 12D *What part of the tongue does it supply? and what type of sensation?*

Taste from the anterior two-thirds of the tongue is carried by some fibres of the **chorda tympani** of the facial nerve (VII) which are distributed to the tongue with the lingual nerve (the chorda tympani also carries preganglionic parasympathetic fibres to the submandibular ganglion.) In the infratemporal fossa (**6.12.4**) identify the chorda tympani as it joins the lingual nerve; trace it backward and upward to the squamotympanic fissure through which it gains the middle ear (p. 97, 100).

Now trace the prominent **inferior alveolar nerve** which arises from the mandibular nerve and lies immediately posterior to the lingual nerve. It passes downward over medial pterygoid to enter the inferior alveolar canal and supply the teeth and gums of the lower jaw and skin of the chin via a branch which passes through the mental foramen. Feel your own mental foramen which lies about a finger's breadth lateral to the midline.

Motor fibres of the mandibular nerve supply muscles of mastication in the infratemporal fossa. A few pass with the inferior alveolar nerve which, before it enters the mandibular canal, gives a small branch — the nerve to mylohyoid — which passes deep to the mandible on the

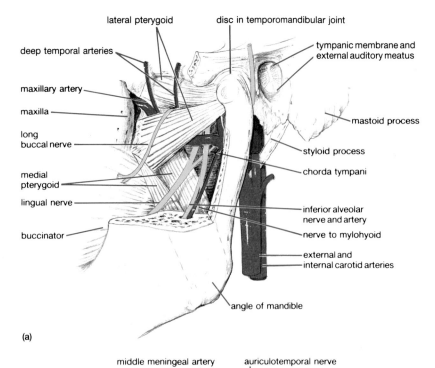

(a)

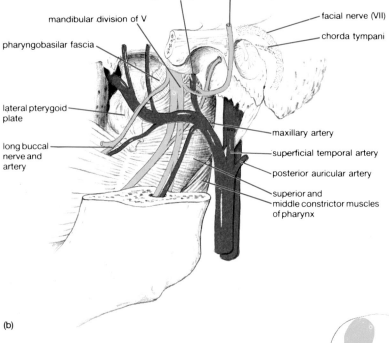

(b)

6.12.4
Mandibular nerve and chorda tympani in
infratemporal fossa.

inferior surface of mylohyoid which it supplies
with the anterior belly of digastric. The mandibular nerve also supplies tensor tympani and
tensor palati.

Beneath the foramen ovale, the otic ganglion,
which cannot be seen with the naked eye, lies
virtually within the mandibular nerve. Only parasympathetic fibres (derived from the tympanic
branch of the glossopharyngeal nerve (IX))
synapse in the ganglion. Its secretomotor fibres
join the auriculotemporal nerve for distribution
to the parotid gland.

In summary, the mandibular division of the
trigeminal nerve supplies **sensory** innervation
to most of the skin over the lower jaw below a
line running from the outer angle of the mouth
to the outer angle of the eye (but excluding the
angle of the jaw), and structures of the face
lying deep to this area (i.e. tongue, teeth, gums,
salivary glands). It also supplies somatic **motor**
innervation to the muscles of mastication, to
tensor tympani, and tensor palati.

Qu. 12E　*Apart from the lower teeth and gums,
what other area is frequently anaesthetized
when the inferior alveolar nerve is infiltrated by
local anaesthetic in the region of the retromolar
fossa in preparation for dental surgery?*

Facial nerve (VII) (6.11.1, 6.12.6, 6.12.7)

The facial nerve is a mixed nerve. The cell
bodies of its motor neurons, which supply both
muscle and glands, lie in the pontine part of the
brainstem, while cell bodies of its sensory fibres
are found in the geniculate ganglion within the
petrous temporal bone. On a prosection of the
interior of the skull, trace the facial nerve and
closely associated vestibulocochlear nerve (VIII)
from where they leave the brainstem between
the pons and medulla, cross the cerebellopontine angle of the subarachnoid space, and
pass beneath the tentorium cerebelli to enter the
internal auditory meatus.

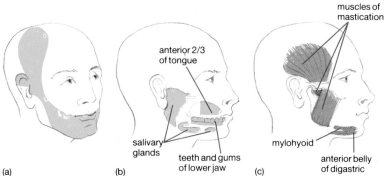

(a)　(b)　(c)

6.12.5
Distribution of mandibular division of V: (a)
sensory to face; (b) sensory to mouth and salivary
glands; (c) motor to muscles of mastication.

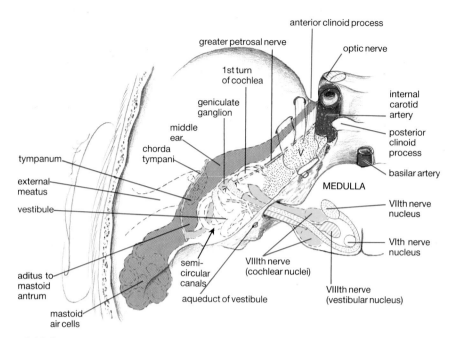

6.12.6
Facial and vestibulocochlear nerves within skull.

On a prosection of the petrous temporal bone, follow the VII nerve laterally through the internal acoustic meatus and petrous temporal bone to the medial wall of the middle ear. Here the nerve is enlarged by the **geniculate** (*genu* = knee) sensory **ganglion** at which the nerve turns at right angles in a bony canal directed backward along the medial wall of the middle ear to its posterior wall. It then makes another right-angled bend and passes downward to leave the skull through the stylomastoid foramen.

Small preganglionic parasympathetic branches (greater superficial petrosal nerve) are given off at the geniculate ganglion. These pierce the thin roof of the petrous temporal bone to reach the middle cranial fossa, then pass through the pterygoid canal to reach the **pterygopalatine ganglion**, and synapse on to postganglionic neurons. In the pterygoid canal the parasympathetic fibres are joined by postganglionic sympathetic fibres from the carotid plexus.

Qu. 12F *What do the postganglionic parasympathetic fibres which leave the pterygopalatine ganglion supply?*

In the posterior wall of the middle ear the facial nerve gives off two branches, the **nerve to stapedius** which carries motor fibres, and the **chorda tympani** which carries both sensory and secretomotor nerves. The chorda tympani passes on to the lateral wall of the middle ear where it runs forward beneath the mucosa on the inner aspect of the tympanic membrane, over the handle of the malleus. It leaves the middle ear through a fissure between the tympanic and petrous parts of the temporal bone (petrotympanic fissure) which opens on to the infratemporal fossa immediately behind the temporomandibular

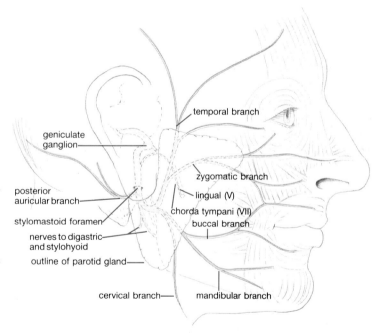

6.12.7
Extracranial course and distribution of facial nerve.

joint, where it joins the lingual nerve (**6.12.4**, **6.12.6**). Its sensory (taste) fibres are distributed to the anterior two-thirds of the tongue; its parasympathetic fibres synapse in the submandibular ganglion which is attached below the lingual nerve on the side of the tongue, and postganglionic secretomotor fibres emerge to supply the submandibular and sublingual salivary glands.

On a prosection, locate the facial nerve as it leaves the stylomastoid foramen; at this point it is entirely motor to muscles. Immediately outside the foramen it supplies stylohyoid, the posterior belly of digastric, and the occipital belly of occipitofrontalis. Follow the nerve into the parotid gland where it divides into its terminal branches (temporal, zygomatic, buccal, mandibular, and cervical) which spread out in a fan-like pattern on the side of the face to supply the muscles of the face, including buccinator and platysma.

Qu. 12G *From which branchial arch are all these muscles derived?*

Note that the cervical branch, which supplies the muscles which control the angle of the mouth, loops down over the submandibular region below the angle of the jaw. Surgical operations (e.g. biopsy of lymph nodes) in this region are relatively common and the nerve must be carefully preserved.

Qu. 12H *The facial nerve can be damaged by virus-induced swelling in the stylomastoid canal. What symptoms comprise the Bell's palsy (6.12.8) that results?*

Qu. 12I *What additional symptoms would you expect if the facial nerve were damaged in the internal acoustic meatus, for example by a tumour?*

Vestibulocochlear nerve (VIII) (6.12.6)

The vestibulocochlear nerve is primarily the sensory nerve from the inner ear and is formed of fibres from the cochlea and vestibular branches from the semicircular canals, utricle, and saccule. Find it as it emerges from the internal acoustic meatus and trace it as it crosses the cerebello-pontine angle, with the facial nerve and labyrinthine artery. It enters the brainstem at the junction of pons and medulla immediately lateral to the facial nerve to synapse in the vestibular and cochlear nuclei of the medulla.

Qu. 12J *Where do the cell bodies of vestibulocochlear neurons lie?*

In addition to vestibular and cochlear sensory fibres, the vestibulocochlear nerve also carries a small bundle of motor fibres from the medulla which end in the cochlea where they modify the process of sound transduction.

C. Radiology

Examine the high-resolution CTs (**6.12.9**) of the petrous temporal bone which show progressively more posterior coronal planes. Identify the main features and trace the course of the facial nerve from the internal auditory meatus, through the geniculate ganglion and middle ear, to its exit from the skull at the stylomastoid foramen. On **6.12.9a** note the external auditory meatus (E), middle ear (M), and epitympanic recess (ER) which is roofed by the thin tegmen tympani and in which is the head of the malleus (H). In the medial wall the canal for the facial nerve (F) and its geniculate ganglion (G) can be seen. Note also the cochlea (C) and carotid canal (CC) beneath the thin floor of the middle ear. In **6.9.12b** the head of the malleus and incus (M, I) are seen, with the facial nerve in its canal running posteriorly on the medial wall of the middle ear. In **6.9.12c** the facial canal is visible just above the oval window (OW) and promontory (P) overlying the first turn of the cochlea (ITC); note also the internal acoustic meatus (I). In

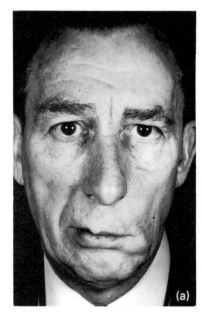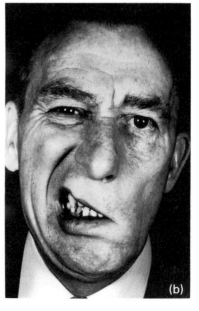

6.12.8
Facial nerve palsy. (a) Patient at rest and (b) when asked to show his teeth.

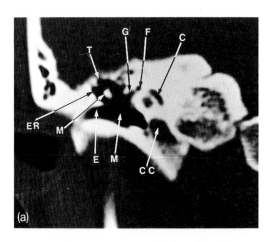

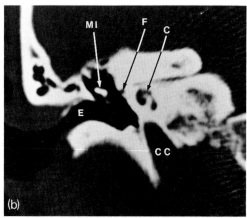

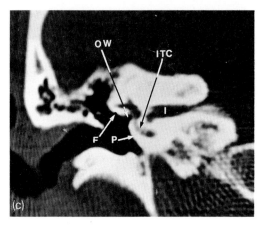

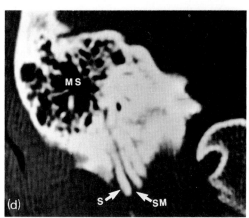

6.12.9
High-resolution coronal CTs through petrous temporal bone.
(a)–(d) show progressively more posterior sections (see text).

6.12.9d the facial canal is seen to descend through the mastoid process and end at the stylomastoid foramen (SM) medial to the styloid process (S); the mastoid air cells can also be seen.

Requirements:

Skull with removable skull cap.

Prosections of the maxillary nerve within the skull, in the pterygomaxillary fissure and posterior aspect of the upper jaw; the lateral aspect of the nose to show the pterygopalatine ganglion and its associated nerves; the infraorbital nerve and zygomatic branches in the face.

Prosections of the mandibular nerve in the infratemporal fossa, on the side of the tongue, terminal cutaneous branches in the face.

Prosections of the facial nerve from brainstem to the internal auditory meatus in the posterior cranial fossa, in the middle ear, in the parotid gland, and on the face.

Prosections of the vestibulocochlear nerve in the posterior cranial fossa.

CTs and MRIs of the petrous temporal bone showing the course of the facial nerve.

Seminar 13

Cranial nerves; glossopharyngeal, vagus, accessory, hypoglossal

The **aim** of this seminar is to study the function, origin, course, and distribution of the **glossopharyngeal** (IX), **vagus** (X), **accessory** (XI), and **hypoglossal** (XII) nerves. In the head and neck, branches of the IX, X, and the cranial root of XI nerves supply the muscles of the palate, pharynx, and larynx on which swallowing and phonation depend, and also supply sensation to these regions and to the ear. The spinal root of XI supplies sternomastoid and trapezius; and the hypoglossal nerve supplies virtually all the muscles of the tongue.

A. Living anatomy

An immediate test of the integrity of the glossopharyngeal, vagus, and cranial accessory

6.13.1
Glossopharyngeal (IX) and hypoglossal (XII) nerves.

nerves is to listen to the quality of the voice and to enquire about the ability to swallow without food 'going down the wrong way' or entering the nasopharynx. To test the glossopharyngeal nerve, lightly touch the posterior third of the tongue or wall of the oropharynx; this should induce a feeling of 'gagging'. To test the motor components of IX, X, and cranial XI, ask the subject to say 'Aah' and establish that the soft palate is elevated in the midline, and ask the subject to swallow some water and observe both the swallowing and the elevation of the larynx that should accompany it (pharyngeal plexus); check that the pitch of the voice can be varied (vagus nerve, external laryngeal branch). To check the movements of the vocal folds (vagus nerve, recurrent laryngeal branch) it is necessary to use a laryngoscope.

The spinal accessory (XI) nerve is vulnerable to penetrating injuries as it crosses the posterior triangle of the neck just beneath its roof. Mark its course by drawing a line from a point one-third of the way down the posterior border of sternomastoid to another two-thirds down the anterior border of trapezius.

Qu. 13A *How would you test the integrity of the spinal accessory nerve?*

B. Prosections

Glossopharyngeal nerve (IX) (6.13.1)

The glossopharyngeal nerve is mixed and supplies motor and sensory fibres to derivatives of the 3rd branchial arch, and secretomotor fibres to the parotid gland and vestibule of the mouth, via the otic ganglion. Cell bodies of sensory components are situated in ganglia lying immediately outside the skull whereas cell bodies of motor neurons are situated in the medulla.

Identify the glossopharyngeal nerve as it emerges from the brainstem as the most cranial of the rootlets between the pyramid and olive, and leaves the skull through the anterior part of the jugular foramen. Follow it as it passes downward through the neck and forward between the internal jugular vein and internal carotid artery, medial to the styloid process, to reach the pharynx. It then passes into the pharynx with stylopharyngeus and the pharyngeal branch of the vagus nerve, between the superior and middle constrictors.

The glossopharyngeal nerve gives a small but important branch, the **sinus nerve**, which supplies the baroreceptors and chemoreceptors of the carotid sinus and carotid body. It supplies **stylopharyngeus** which arises from the styloid process and passes downward and forward to merge with the pharyngeal constrictors. **Lingual** branches supply the posterior third of the tongue with both taste and common sensation; **pharyngeal** branches join the **pharyngeal plexus** to supply sensation to the oropharynx; and **tonsillar** branches supply the tonsillar fossa, pillars, and fauces and soft palate.

The glossopharyngeal nerve also gives a **tympanic branch** which carries secretomotor fibres to the parotid gland. It leaves the main nerve just below the jugular foramen and passes through a small bony canal to reach the middle ear where it forms the **tympanic plexus** which supplies sensation to the middle ear. The secretomotor fibres form the **lesser (superficial) petrosal nerve** which passes through the tegmen tympani into the middle cranial fossa and leaves the skull through the foramen ovale to synapse in the otic ganglion.

Qu. 13B *Through which peripheral nerve are they distributed to the parotid gland?*

Vagus nerve (X) (6.13.2)

The vagus nerve, like the glossopharyngeal, has several different motor and sensory components which supply the derivatives of the 4th and more caudal branchial arches, the heart, the foregut and midgut, and their derivatives. Cell bodies of its motor neurons (to both muscle and glands) lie in the medulla while those of its sensory neurons lie in two ganglia immediately outside the skull. It is predominantly a **sensory** nerve, comprising 70 per cent sensory fibres and 30 per cent motor fibres.

Identify the vagus nerve as it emerges from the medulla, just caudal to the glossopharyngeal nerve, as a series of rootlets which unite and pass through the jugular foramen, where it is joined by the **cranial root of the accessory nerve** (see below) which carries motor fibres to the (branchial) muscles of the palate, pharynx, and larynx. Trace the main trunk of the vagus as it runs vertically downward in the neck, lying posteriorly in the carotid sheath between the internal jugular vein and the internal carotid or common carotid artery, to pass into the thorax (Vol. 2, Chapter 5, Seminar 5).

From the superior ganglion (which contains the cell bodies of somatic afferent fibres) arises a **meningeal** branch which supplies the posterior cranial fossa and a small **auricular** branch which pierces the lateral wall of the jugular foramen to supply sensory fibres to the posterior inferior part of the external auditory meatus and tympanic membrane (and the cranial surface of the pinna).

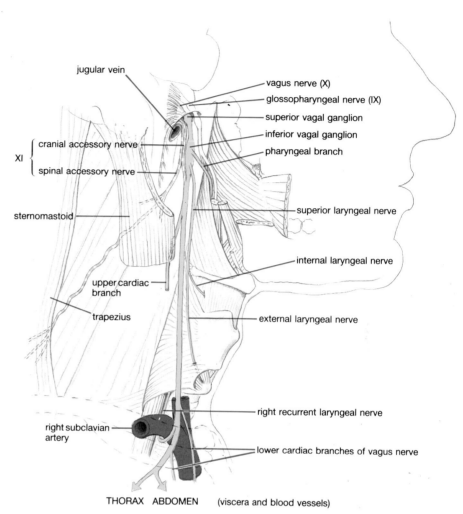

6.13.2
Vagus (X) and accessory (XI) nerves.

Qu. 13C *What effect might stimulation of the external auditory meatus have on heart rate?*

Identify branches of the vagus nerve in the neck. A **pharyngeal** branch runs with the glossopharyngeal nerve to reach the pharynx and join the **pharyngeal plexus** to which it contributes sensory fibres to the laryngopharynx, and motor fibres to muscles of the pharynx and palate (except stylopharyngeus and tensor palati). The **superior laryngeal nerve** runs behind the carotid sheath to the larynx; its **internal** and **external laryngeal** branches supply, respectively, sensation to the valleculae and larynx above the vocal folds and motor fibres to cricothyroid.

Qu. 13D *How does the internal laryngeal nerve enter the larynx?*

The **recurrent laryngeal nerve** is given off in the thorax (Vol. 2, Chapter 5, Seminar 5), re-enters the neck, and runs upward between the oesophagus and trachea and among the branches of the inferior thyroid artery to enter the larynx where it supplies sensation to the vocal folds and the airway below, and motor fibres to the intrinsic muscles of the larynx (except for cricothyroid).

Qu. 13E *On the left side the recurrent laryngeal nerve loops beneath the ductus arteriosus; beneath which vessel does it loop on the right?*

Qu. 13F *What would result if the left recurrent laryngeal nerve were stretched by an aneurysm of the arch of the aorta?*

Cardiac branches of the vagus arise in the neck, reflecting the more cranial position of the heart during development. A chemoreceptor branch (aortic nerve) supplies the aortic arch, and superior and inferior cardiac branches run with the vagus nerve to supply parasympathetic and sensory fibres to the heart. Within the thorax the vagus nerve supplies visceral motor and sensory fibres to the oesophagus, trachea, and lungs, and in the abdomen it supplies abdominal viscera derived from the fore- and midgut (stomach, pancreas, liver, and gall bladder, small and large intestine as far as the splenic flexure of the colon) (Vol. 2, Chapter 6, Seminar 5).

Accessory nerve (XI) (6.13.2)

The **cranial root of the accessory** nerve contains motor fibres from cells of the nucleus ambiguus of the medulla which supply striated muscles of the palate (except tensor palati), pharynx (except stylopharyngeus), and larynx. Identify its rootlets as they emerge from the lateral aspect of the brainstem below those of the glossopharyngeal and vagus nerves and join to form a trunk which runs toward the jugular foramen. This trunk joins the spinal root of the accessory nerve for a short distance, but soon diverges from it to join the vagus nerve below the jugular foramen. The cranial accessory fibres are distributed with the pharyngeal and laryngeal branches of the vagus.

Identify the **spinal root of the accessory nerve**. This motor nerve arises from cell bodies in the anterior horn of the cervical (C2–4) part of the spinal cord. Its rootlets leave the cord, collect together and pass upward through the foramen magnum; they briefly join the cranial root of the accessory nerve and leave the skull through the jugular foramen. Outside the skull, the spinal part of the accessory nerve separates from the cranial part, and passes laterally into sternomastoid which it supplies. It emerges from

sternomastoid, and crosses the posterior triangle of the neck beneath the roof of investing cervical fascia, to supply trapezius.

Qu. 13G *If the accessory nerve were damaged on one side (6.13.3) by a fracture through the base of the skull, what would you expect to find:*
  (a) when testing the movements of the soft palate?
  (b) when asking the patient to shrug against resistance?
  (c) when asking the patient to turn his chin upward towards the opposite side?

Hypoglossal nerve (XII) (6.13.1)

The hypoglossal nerve is motor and sensory (proprioceptive) to the muscles of the tongue. Its motor cell bodies lie close to the midline of the caudal medulla. Identify its rootlets as they emerge from the ventral aspect of the medulla (between olive and pyramid) and follow the nerve through the subarachnoid space to the hypoglossal canal. The hypoglossal nerve emerges from the canal into the neck behind the carotid sheath, spirals downward and posteriorly to pass between the internal carotid artery and internal jugular vein. Locate the nerve at the level of the styloid process, where it turns forward and medially to form a large loop which crosses the internal and external carotid arteries and the loop of the lingual artery to lie on hyoglossus under cover of mylohyoid. Its branches ramify within the tongue, supplying all its muscles except palatoglossus.

Qu. 13H *Describe the result of damage to the right hypoglossal nerve within the skull (see 6.13.4).*

As it leaves the skull and crosses the transverse process of the atlas the hypoglossal nerve is joined by fibres from the anterior primary ramus of **C1**; these are distributed via the **ansa cervicalis** (see p. 129).

Requirements:
  Skull with removable skull cap.
  Prosections of the brainstem and upper cervical spinal cord *in situ* with the laminae of the cervical vertebrae removed, to show the intracranial course of the glossopharyngeal, vagus, accessory, and hypoglossal nerves; the neck to show the course and distribution of the glossopharyngeal, vagus, accessory, and hypoglossal nerves and their branches; also the ansa cervicalis.

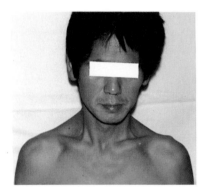

6.13.3
Left spinal accessory nerve (XI) palsy.

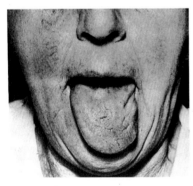

6.13.4
Left hypoglossal nerve (XII) palsy.

Seminar 14

Cervical somatic and sympathetic nerves

The **aim** of the seminar is to study the cervical spinal nerves and the plexuses that they form, the cervical sympathetic chain, and the root of the neck, in the living and with the use of prosections.

A. Living anatomy

Define the posterior border of sternomastoid, the anterior border of trapezius, and the clavicle, and draw on your partner's neck the outline of the posterior triangle of the neck. Palpate the muscles of the floor of the triangle which are covered by the prevertebral fascia; the most prominent is scalenus medius. Run your finger firmly across the lower part of the posterior triangle to feel the trunks of the brachial plexus (formed from the lower cervical nerves) which pass laterally to disappear beneath the clavicle and enter the axilla (Vol. 1, Chapter 6, Seminar 8). The other branches of the cervical plexus are too small to feel but, after you have studied the prosections of the cervical plexus, mark the course of the main branches on your partner's neck. The first cervical nerve does not innervate any skin, but C2, C3, and C4 supply bands of skin around the neck and down to the manubriosternal joint.

Look at the skeleton and remind yourself of the obliquity of the 1st rib. Mark on to your partner's neck the apex of the lung which extends upward for 2–3 cm behind the medial third of the clavicle, but reaches no higher than the neck of the 1st rib.

B. Prosections of the cervical and brachial plexuses and cervical sympathetic chain

Cervical plexus (6.14.1)

On a prosection locate scalenus anterior and the nerves that emerge from its lateral border (**6.14.3**). The cervical plexus is formed from the **anterior primary rami** of the **upper four cervical nerves**; it lies posterior to the carotid sheath and its contents, and anterior to scalenus medius. C1 emerges above the atlas while C2, C3, and C4 traverse the intervertebral foramina above the 2nd, 3rd, and 4th cervical vertebrae.

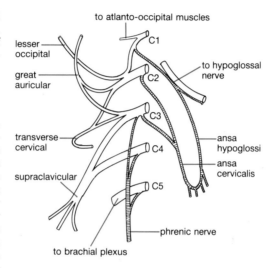

6.14.1
Cervical plexus.

The anterior primary rami of all the cervical spinal nerves supply the prevertebral muscles segmentally. C1 also gives a branch which joins the hypoglossal nerve (XII) to supply thyrohyoid and geniohyoid, and to provide the upper limb of the ansa (C1) which joins the lower limb (C2,3) to form the **ansa cervicalis** which supplies the infrahyoid strap muscles.

The superficial sensory branches derived from the anterior primary rami of C2, C3, and C4 pierce both the prevertebral and investing layer of cervical fascia at the posterior border of sternomastoid to reach the skin (**6.14.2**). C1 does not supply skin.

Locate:

● The **lesser occipital nerve** (C2) which ascends along the posterior border of the sternomastoid to supply an area of scalp behind the pinna and the upper part of the cranial surface of the pinna.

● The **great auricular nerve** (C2, 3) which ascends across sternomastoid and divides to supply skin over the parotid gland and the angle of the jaw, skin over the mastoid process, and the lobe and lower part of the cranial surface of the pinna.

● The **transverse cutaneous nerve** (C2, 3)

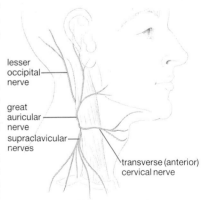

6.14.2
Cutaneous branches of cervical plexus.

which runs forward over sternomastoid to supply a relatively large area of skin over the side and front of the neck.

• The **supraclavicular nerves** (C3, 4) which arise from a common trunk, emerge from the posterior border of sternomastoid, and descend to supply the skin over the lower part of the neck and over the clavicle as far as the 2nd intercostal space. The most lateral supplies the tip of the shoulder. This area is not infrequently the site of pain referred from the region of the diaphragm (e.g. from inflammation of the gall bladder); afferent impulses in the phrenic nerve (C4) enter the spinal cord and are perceived as pain in the territory supplied by the lateral branch of the supraclavicular nerve (C4).

The **phrenic nerve** (C3, 4, 5) emerges from the posterior border of scalenus anterior and runs vertically downward on this muscle beneath the prevertebral fascia to enter the thorax through which it passes to supply motor fibres to the diaphragm and sensory fibres to the pleura and peritoneum which cover the diaphragm (see Vol. 2, Chapter 5, Seminar 2).

Qu. 14A *What would be the result of damage to the spinal cord at C3 and above?*

The **posterior primary rami** of cervical nerves C2–8 supply skin over the back of the head and neck.

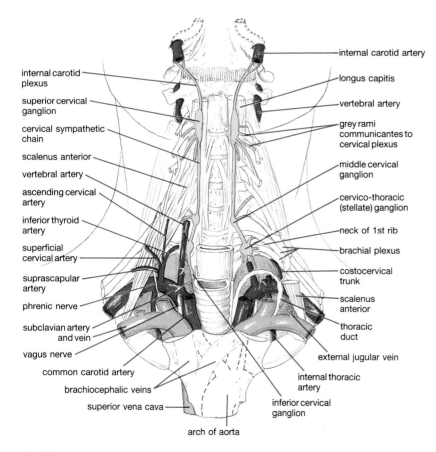

internal carotid plexus

superior cervical ganglion

cervical sympathetic chain

scalenus anterior

vertebral artery

ascending cervical artery

inferior thyroid artery

superficial cervical artery

suprascapular artery

phrenic nerve

subclavian artery and vein

vagus nerve

common carotid artery

brachiocephalic veins

superior vena cava

arch of aorta

internal carotid artery

longus capitis

vertebral artery

grey rami communicantes to cervical plexus

middle cervical ganglion

cervico-thoracic (stellate) ganglion

neck of 1st rib

brachial plexus

costocervical trunk

scalenus anterior

thoracic duct

external jugular vein

internal thoracic artery

inferior cervical ganglion

6.14.3
Deep aspect of neck and sympathetic chain.

Brachial plexus (6.14.3)

The anterior primary rami of the roots of C5 to T1 form the brachial plexus which supplies the upper limb. Explore the space between the scalenus anterior and scalenus medius and locate the separate **nerve roots**; those of C5 and C6 unite to form the **upper trunk**; C7 forms the **middle trunk**; and C8 and T1 unite to form the **lower trunk**. These trunks pass laterally through the lower part of the posterior triangle behind the prevertebral fascia, and leave the neck by passing behind the clavicle to enter the axilla, ensheathed by a lateral extension of the fascia. The lower trunk lies on the 1st rib, immediately behind the subclavian artery. (Occasionally the brachial plexus arises from roots C4–8 ('prefixed') or C6–T2 ('postfixed'). The lowest root (T2) of a postfixed plexus is liable to trauma as a result of its longer course from the thorax to the axilla.)

A number of branches arise from the brachial plexus in the neck which supply muscles of the upper limb: a branch from C5 passes back through scalenus medius to supply the rhomboid muscles; branches from C5 and C6 unite to supply subclavius; the **suprascapular nerve** (C5, 6) runs backward and passes through the suprascapular notch to supply supraspinatus, infraspinatus, and the shoulder joint; and branches from the C5, 6, and 7 roots unite behind the plexus to form the **long thoracic nerve** which crosses the first rib to supply serratus anterior on the medial wall of the axilla (Vol. 1, Chapter 6, Seminar 8).

Cervical sympathetic chain (6.14.3)

On a prosection locate the **cervical** part of the **sympathetic trunk** which lies on the prevertebral fascia dorsal to the carotid sheath and its contents. It is characterized by three ganglia. Identify the large, elongated **superior cervical ganglion** which lies on the ventral aspect of the transverse processes of the 2nd and 3rd cervical vertebrae, the small **middle cervical ganglion** (if present), and the **inferior cervical ganglion** which is often fused with the 1st thoracic sympathetic ganglion to form a **stellate ganglion** on the ventral aspect of the neck of the first rib. The cells of the ganglia give postganglionic fibres which travel to skin and muscles of the head, neck, and upper limbs via the cervical spinal nerves, and reach the organs and internal structures of the head and neck largely along the appropriate blood vessels. They supply sweat glands and arrector pili muscles, smooth muscle of blood vessels, the lacrimal and salivary glands, dilator pupillae, and some fibres of levator palpebrae. In addition, sympathetic branches to the heart, like those of the vagus, arise in the neck and pass down through the neck into the thorax to supply the heart. A plexus on the internal carotid artery arises from the superior cervical ganglion; it supplies sympathetic fibres to the ciliary ganglion and the pterygopalatine ganglion and to branches of the artery.

Qu. 14B *By what route do the submandibular and otic ganglia receive sympathetic postganglionic fibres?*

The preganglionic fibres which innervate the cervical ganglion cells arise from neurons which lie, not in the cervical spinal cord, but in the lateral horn of the upper four or five *thoracic* segments of the cord. They pass into the thoracic sympathetic chain which ascends over the neck of the 1st rib to become the cervical sympathetic trunk. Therefore, all sympathetic innervation to the head and neck is derived from within the thorax.

Qu. 14C *Damage to the sympathetic chain by a tumour of the apex of the lung leads to the development of a 'Horner's syndrome' (6.14.4) in which the upper eyelid droops (partial ptosis), the skin of the face is dry and flushed, the pupil is constricted, and the eyeball appears sunken (enophthalmos). Account for these signs.*

Root of the neck (6.14.3, 6.14.5)

The root of the neck is a narrow region through which structures pass between the neck and the thorax, and which it is appropriate to review at this point. On an articulated skeleton, note the bony margins of the thoracic inlet: the manubrium, the obliquely-set 1st rib, and the first thoracic vertebra. On a prosection of the root of the neck and upper part of the thorax note that, in the midline, the space between the manubrium and the first thoracic vertebra is almost entirely taken up by the oesophagus and trachea. The trachea passes downward and slightly to the right to divide into the two main bronchi at the level of the 2nd costal cartilage. Identify the large arteries arising from the aortic arch. The brachiocephalic trunk divides to form the right subclavian and common carotid arteries behind the right sternoclavicular joint, while the left common carotid and subclavian arteries arise directly from the arch of the aorta. As they pass upward into the neck, the brachiocephalic artery and left common carotid artery diverge to lie on either side of the trachea. Follow the subclavian arteries across the 1st rib where they lie behind the insertion of scalenus anterior before entering the axilla.

Qu. 14D *What branches arise from the subclavian artery?*

Trace the subclavian veins as they leave the axilla and cross the 1st rib anterior to scalenus anterior to enter the thorax. Each receives the external jugular vein, then joins with the corresponding internal jugular vein to form a brachiocephalic vein. The right brachiocephalic vein passes downward into the thorax behind the right manubriosternal joint and is joined by the left brachiocephalic vein which passes diagonally from left to right immediately deep to the manubrium to form the superior vena cava. In a young child, the left brachiocephalic vein lies at a higher level, above the upper border of the manubrium,

6.14.4
Horner's syndrome (arrow).

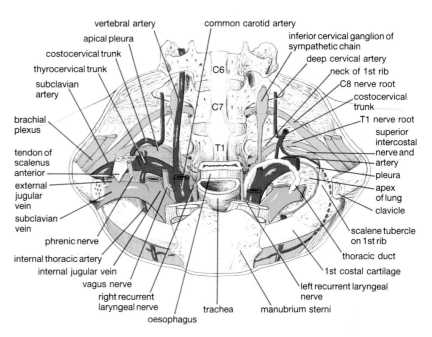

6.14.5
Root of neck.

where it is at risk from penetrating injuries, and emergency tracheostomy (opening the trachea in cases of upper airway obstruction).

Locate the **thoracic duct** which emerges from the thorax to the left of the oesophagus and drains lymph from the thorax, abdomen, and lower limbs. It passes laterally, behind the carotid sheath, and is usually joined by a **subclavian trunk** draining the upper limb and a **jugular trunk** draining the deep cervical chain of nodes, before it drains into the bloodstream at the junction of the left internal jugular and subclavian veins. In the equivalent position on the right side you may be able to find a single, short **right lymphatic duct**, or separate lymphatic trunks draining the right side of the neck, upper limb, thoracic wall, and mediastinum into the venous circulation.

Note the relationships of the nerves which pass between the neck and the thorax. Follow each vagus nerve (X) from the posterior part of the carotid sheath into the thorax: on the right the nerve descends posterior to the internal jugular vein and crosses anterior to the first part of the subclavian artery to enter the thorax where it lies behind the right brachiocephalic vein; on the left, the vagus enters the thorax behind the left brachiocephalic vein and passes down between the left common carotid and subclavian arteries and to the left side of the aortic arch. Locate the right and left recurrent laryngeal branches of the vagus nerves; the right loops under the right subclavian artery, the left beneath the ligamentum arteriosum and arch of the aorta; trace them upward between the oesophagus and trachea.

Follow the phrenic nerve as it passes from the cervical plexus vertically downward on scalenus anterior beneath the prevertebral fascia to enter the thorax between the subclavian artery and vein.

Examine a prosection of the apical regions of the thorax from which the lungs have been removed. Identify the suprapleural membrane and parietal pleura which form the domed roof over each lung apex and, within the thorax, locate the trachea, oesophagus, vagus, and phrenic nerves, and the great vessels. Locate the root of T1 as it emerges from the intervertebral space between the 1st and 2nd thoracic vertebrae and crosses the 1st rib posterior to the subclavian artery to leave the thorax and join the brachial plexus. Follow the costocervical trunk as it arches over the apex of the lung and divides to give the deep cervical artery and (medial to T1 on the neck of the 1st rib) the superior intercostal artery which supplies the upper two intercostal spaces. Medial to the superior intercostal artery, the sympathetic chain ascends on the neck of the 1st rib to enter the neck.

Qu. 14E *What neurological abnormalities other than a Horner's syndrome might a patient with a large tumour of the apex of the lung exhibit?*

C. Radiology

Qu. 14F *What anomaly is arrowed in radiograph* **6.14.6**; *which nerve is it most likely to affect; and what symptoms would result from damage to the nerve?*

Requirements:

Articulated skeleton.

Prosections of the neck to show the cervical plexus and its branches, and the roots and trunks of the brachial plexus; the root of the neck and the superior mediastinum with the manubrium sternum removed; the apices of the thorax with the lungs removed to show the sympathetic chain and stellate ganglion, superior intercostal artery, and T1 nerve root.

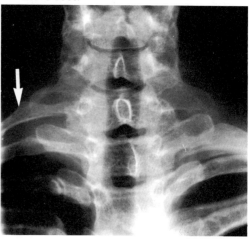

6.14.6

See **Qu. 14F.**

Seminar 15

Cross-sectional anatomy of the head and neck revealed by CT

The **aim** of this seminar is to relate the three-dimensional morphology of the head and neck to its appearance in transverse sections and sections in other planes, as revealed by computed tomography (CT).

Computed tomography and magnetic resonance imaging are non-invasive techniques which are already of considerable diagnostic importance and which are likely to become even more frequently used in future. It is therefore important that medical students should be able to recognize on CTs the normal shape, position, and relations of all the major structures in the head and neck. Once this has been accomplished, it should not be long before lesser structures and abnormalities can be recognized.

Place a cover across the adjacent labelled diagrams and discover how many structures you can identify before checking your observations. Always remember the orientation of these images. It is as if you are you were within the cavity of the body, looking rostrally at the head and neck.

Figure 6.15.1 (a–e) are transverse sections (the head being positioned so that the outer angle of the eye and the external auditory meatus are in the same horizontal plane) through the cranial cavity, taken at various levels (6.15.1f).

Figure 6.15.2 (a–c) are sections, taken in the horizontal plane but with the chin tilted downward by 10°, of the cranial cavity which emphasize features of the middle cranial fossa (6.15.2d). Both sets of images have been taken after contrast material has been introduced into the cerebrospinal fluid of the subarachnoid space by lumbar puncture. This technique obliterates details of the brain and brainstem and thus enables the bony structures to be seen more readily against a uniform grey background.

Fig. 6.15.3 (a–g) are horizontal sections taken at progressively lower levels through the face and neck (6.15.3h).

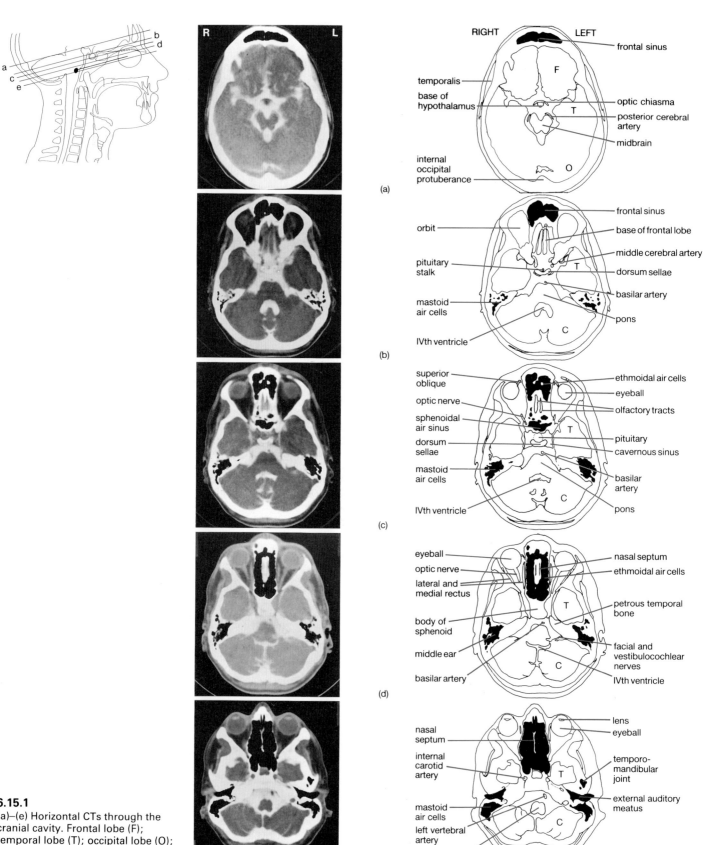

6.15.1

(a)–(e) Horizontal CTs through the cranial cavity. Frontal lobe (F); temporal lobe (T); occipital lobe (O); cerebellum (C).

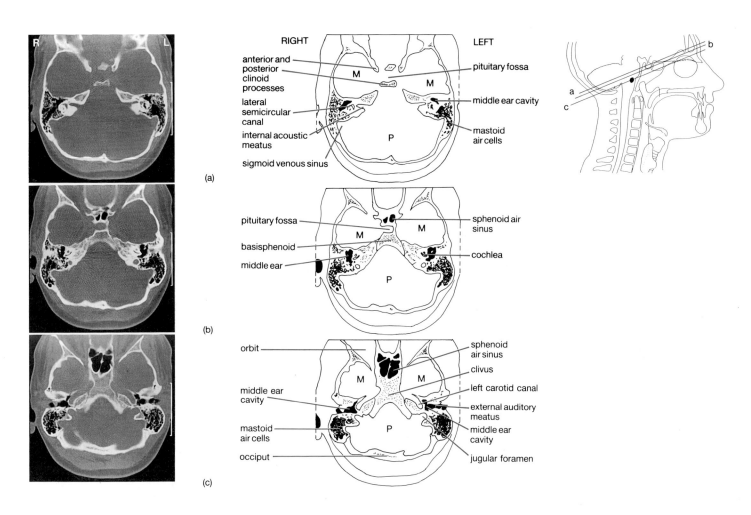

6.15.2
(a)–(c) Horizontal CTs taken with chin 10° down.
Middle cranial fossa (M); posterior cranial fossa
(P).

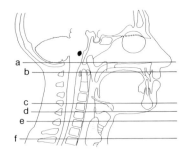

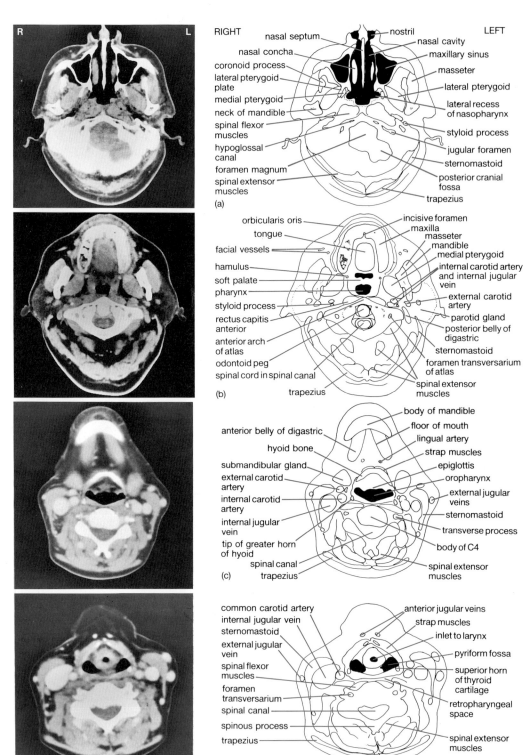

RIGHT   LEFT

(a)
- nasal septum
- nostril
- nasal cavity
- nasal concha
- maxillary sinus
- coronoid process
- masseter
- lateral pterygoid plate
- lateral pterygoid
- medial pterygoid
- lateral recess of nasopharynx
- neck of mandible
- styloid process
- spinal flexor muscles
- jugular foramen
- hypoglossal canal
- sternomastoid
- foramen magnum
- posterior cranial fossa
- spinal extensor muscles
- trapezius

(b)
- orbicularis oris
- incisive foramen
- maxilla
- tongue
- masseter
- facial vessels
- mandible
- medial pterygoid
- hamulus
- internal carotid artery and internal jugular vein
- soft palate
- pharynx
- external carotid artery
- styloid process
- rectus capitis anterior
- parotid gland
- anterior arch of atlas
- posterior belly of digastric
- sternomastoid
- odontoid peg
- foramen transversarium of atlas
- spinal cord in spinal canal
- spinal extensor muscles
- trapezius

(c)
- body of mandible
- floor of mouth
- anterior belly of digastric
- lingual artery
- hyoid bone
- strap muscles
- submandibular gland
- epiglottis
- external carotid artery
- oropharynx
- internal carotid artery
- external jugular veins
- internal jugular vein
- sternomastoid
- transverse process
- tip of greater horn of hyoid
- body of C4
- spinal canal
- spinal extensor muscles
- trapezius

(d)
- common carotid artery
- anterior jugular veins
- internal jugular vein
- strap muscles
- sternomastoid
- inlet to larynx
- external jugular vein
- pyriform fossa
- spinal flexor muscles
- superior horn of thyroid cartilage
- foramen transversarium
- retropharyngeal space
- spinal canal
- spinous process
- spinal extensor muscles
- trapezius

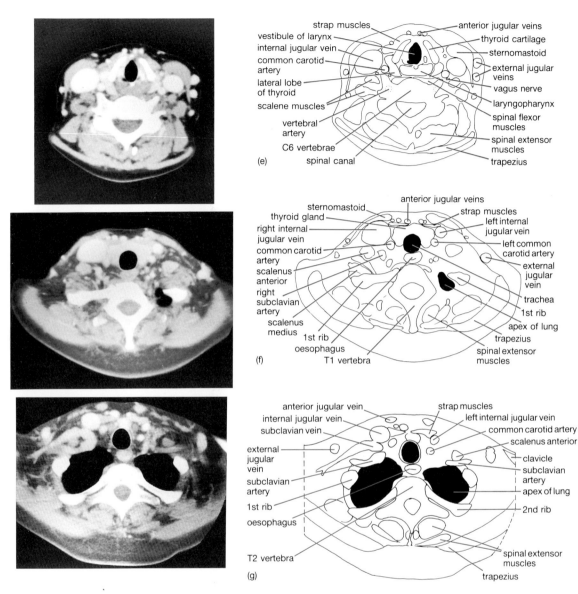

(e)

strap muscles
vestibule of larynx
internal jugular vein
common carotid artery
lateral lobe of thyroid
scalene muscles
vertebral artery
C6 vertebrae
spinal canal
anterior jugular veins
thyroid cartilage
sternomastoid
external jugular veins
vagus nerve
laryngopharynx
spinal flexor muscles
spinal extensor muscles
trapezius

(f)

sternomastoid
thyroid gland
right internal jugular vein
common carotid artery
scalenus anterior
right subclavian artery
scalenus medius
1st rib
oesophagus
T1 vertebra
anterior jugular veins
strap muscles
left internal jugular vein
left common carotid artery
external jugular vein
trachea
1st rib
apex of lung
trapezius
spinal extensor muscles

(g)

anterior jugular vein
internal jugular vein
subclavian vein
external jugular vein
subclavian artery
1st rib
oesophagus
T2 vertebra
strap muscles
left internal jugular vein
common carotid artery
scalenus anterior
clavicle
subclavian artery
apex of lung
2nd rib
spinal extensor muscles
trapezius

6.15.3
(a)–(g) Horizontal CTs through face and neck.

CHAPTER 7

Anatomical basis of some neural reflexes of the head and neck

The **aim** of this chapter is to study a few of the many neural and hormonal reflexes which affect the head and neck. Many involve interaction between the autonomic and somatic nervous systems. The text is purposely brief and emphasizes the basic components of a reflex.

- A **sensory neuron** which transmits information from the external or internal environment to the central nervous system.
- An interconnection between the sensory and motor neurons which usually involves one or more neurons (**interneurons**) within the central nervous system.
- A **motor neuron** which causes a somatic or visceral motor response.

The diagrams in this chapter have been standardized as far as possible, with somatic components coloured blue, sympathetic components yellow, and parasympathetic components green.

The least anatomically complex reflexes involve a direct connection between the afferent and efferent neuron (monosynaptic) as exemplified by the 'stretch reflex' (e.g. the jaw jerk). However, most reflexes also involve one or more central interneurons interposed between the afferent and efferent neuron (polysynaptic). Such interneurons lie at, or in different centres above or below, the level at which the afferent neuron enters the central nervous system. The original sensory stimulus therefore spreads locally and is carried to higher centres, thus enabling the original information to be processed and integrated before the motor (efferent) response

is stimulated. Similarly, by influencing the interneuron pool, descending information from higher centres can have a marked effect on the sensitivity of the reflexes.

The presence and quality of the reflex responses therefore provides an indication of both the integrity of the reflex arc and also the degree of excitation of that arc from other centres.

Jaw jerk reflex (7.1)

If the mouth is held slightly open and the lower jaw tapped sharply from above with a reflex-testing hammer the jaw will close and open again rapidly. This is a monosynaptic 'stretch reflex' involving primarily masseter and temporalis.

- **Afferent**. Stimuli arise from muscle spindles in muscles of mastication that have been stretched and pass via the **mandibular** division of the trigeminal nerve (V) to its proprioceptive nucleus in the midbrain. As the fibres enter the brainstem they give off collaterals which pass directly to the motor nucleus of the trigeminal nerve in the pons.
- **Efferent**. Motor fibres of the **mandibular nerve** stimulate muscles which close the jaw (masseter, temporalis, medial pterygoid).

Light reflex (7.2)

- **Afferent**. Light excites the retinal receptors and is transmitted via the ganglion cells to the

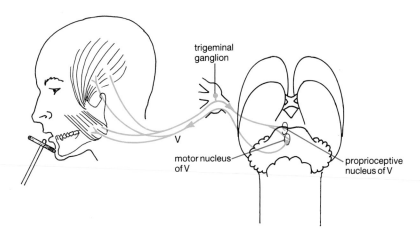

7.1
Jaw jerk.

optic **nerve** and tract. Fibres leave the optic tract (before it reaches the thalamus) to synapse in the **pretectal nucleus**. The information is relayed from the pretectal nucleus by short neurons which synapse bilaterally with parasympathetic neurons in the **(Edinger–Westphal) nucleus of the oculomotor** (III) nerve in the upper part of the midbrain; it is also relayed to the spinal (T1, 2) sympathetic centre controlling dilator pupillae.

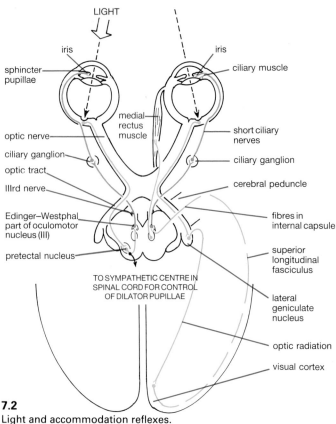

7.2
Light and accommodation reflexes.

• **Efferent** stimuli pass along **parasympathetic fibres of the oculomotor nerve** (III) to the orbit where they synapse in the **ciliary ganglion**. Postganglionic fibres (**short ciliary nerves**) pass to the eyeball to supply sphincter pupillae (which reduces the size of the pupil) and ciliaris (which causes accommodation); inhibition of **sympathetic** tone relaxes dilator pupillae.

This reflex is 'crossed', i.e. both pupils react even if the light stimulus impinges on one eye only. If only the stimulated eye reacts, then the reflex has been interrupted in the brainstem.

Accommodation reflex (7.2)

When focusing on an object a decision is made in the central nervous system concerning the object in the visual field to be focused on. Information from the retina, optic nerve, tract, and radiation are therefore necessary, but are not an afferent limb of a reflex in the sense that this term is commonly used. Focusing on nearby objects causes pupillary constriction in addition to changes in the lens and convergence of the eyes.

The information passes via a series of neurons in the cerebral hemisphere and via the internal capsule of the same side to the nucleus of the oculomotor nerve, and to the spinal (T1, 2) sympathetic centre controlling dilator pupillae.

• **Efferent** stimuli pass in preganglionic parasympathetic fibres of the **oculomotor** (III) nerve to the orbit where they synapse in the **ciliary ganglion**. Postganglionic fibres (**short ciliary nerves**) passing to the eyeball stimulate contraction of sphincter pupillae and the ciliary muscle. Contraction of sphincter pupillae and relaxation of dilator pupillae close the pupil; contraction of the ciliary muscle slackens the ligament of the lens which increases in curvature for near vision; and contraction of the medial,

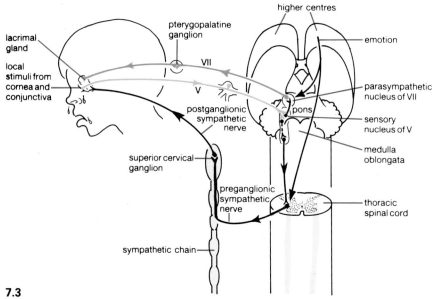

7.3
Lacrimation reflex.

superior, and inferior recti (**oculomotor nerve**) converges the eyes on the near target. The pupillary changes may be secondary to the convergence.

Lacrimation reflex (7.3)

● **Afferent**. This reflex is triggered by many different emotional stimuli or by irritation of the cornea and conjunctiva from which information is carried by the **ophthalmic nerve** (V) to its sensory nucleus in the brainstem.

The stimulus spreads through interneurons to: (a) parasympathetic cells of the 'superior salivatory centre' (VII nerve) in the pons; and (b) preganglionic sympathetic cells in the lateral horn of the upper thoracic spinal cord which synapse on postganglionic cells of the superior cervical sympathetic ganglion.

● **Efferent**. Stimuli pass via (a) **parasympathetic** fibres of the facial nerve which synapse in the pterygopalatine ganglion and (b) postganglionic sympathetic fibres to activate the lacrimal gland. Stimulation of both sympathetic and parasympathetic fibres causes increased lacrimal secretion.

Corneal and 'blink' reflexes (7.4)

● **Afferent**. Reflex closure of the eye is triggered by touching the cornea, by emotion, or by bright light. Stimuli pass respectively from the **ophthalmic nerve** (V), higher centres, or the retina to a centre involved with visual reflexes situated in the midbrain.

The impulse is spread by interneurons which carry the information to the motor nucleus of the facial nerve in the pons.

● **Efferent**. Fibres of the **facial nerve** cause a brief contraction of palpebral fibres of orbicularis oculi. This reflex protects the cornea from damage, protects the retina from lengthy exposure to bright light, and sweeps lacrimal secretion across the surface of the eye which removes harmful particles and moistens the cornea and conjunctiva.

Stapedial reflex (7.5)

● **Afferent**. Reflex contraction of stapedius (and tensor tympani) is triggered both by loud sounds and immediately prior to speech. Stimuli from the **auditory nerve** (VIII) and from higher centres immediately prior to speech spread to the somatic motor nucleus of the facial nerve in the pons.

● **Efferent** stimuli pass in motor axons of the **facial nerve** (VII) to stapedius which, on contraction, reduces movements of stapedius and the ossicle chain of the middle ear, and thus attenuates sound inputs to the oval window.

Tensor tympani also contracts in response to loud noises; the efferent pathway in this case is the mandibular nerve (V).

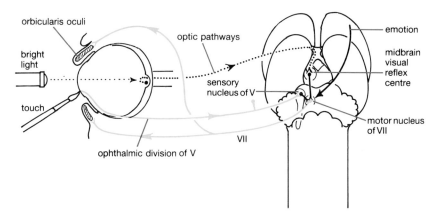

7.4
Blink and corneal reflexes.

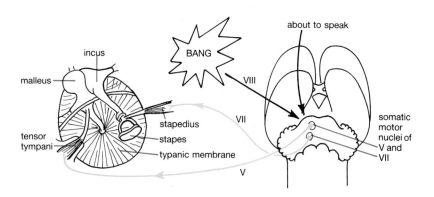

7.5
Stapedial reflex.

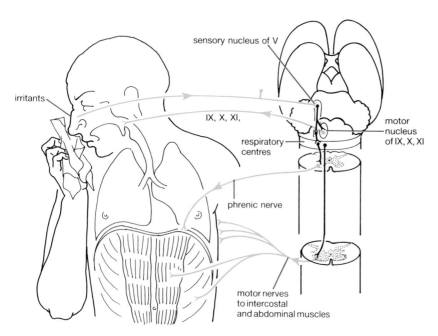

7.6
Sneeze reflex.

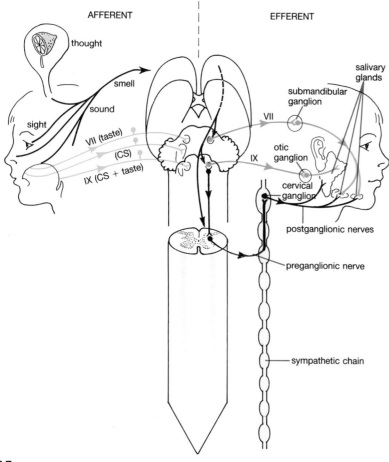

7.7
Salivation reflex.

Sneezing reflex (7.6)

● **Afferent**. Irritants stimulate receptors in the nasal mucous membrane and pass via somatic sensory axons of the **ophthalmic** or **maxillary** divisions of the trigeminal nerve (V) to its sensory nucleus in the brainstem.

The information spreads to respiratory centres and the nucleus ambiguus in the medulla.

● **Efferent**. Somatic motor fibres of the **phrenic nerves** (C3, 4, 5) to the diaphragm and **intercostal nerves** (T1–T12) to the external intercostal muscles elicit a deep inspiration. This is followed by a violent exhalation (intercostal and abdominal wall muscles; **intercostal nerves**) after build-up of pressure against a closed glottis similar to that which occurs in coughing. The stream of air is directed through the nose by closure of the oropharyngeal isthmus (palatoglossus; **pharyngeal plexus**) to expel the irritating particle. The **cough reflex** (Vol. 2, Chapter 7) that is stimulated by irritation of receptors in the larynx and trachea (**laryngeal branches of vagus nerve**) is essentially similar except that the oropharyngeal isthmus remains open.

Salivation reflex (7.7)

● **Afferent**. This reflex is triggered by a sharp taste (lemon juice is very effective) by the sight, and sound, smell, or even thought of food, and by the presence of items in the mouth. These stimuli are carried by taste fibres (chorda tympani and glossopharyngeal nerves) via the solitary nucleus of the medulla, from higher centres in the brain, or by somatic sensory neurons of the tongue and mouth (lingual and glossopharyngeal nerves).

The information spreads via interneurons to (a) preganglionic parasympathetic cells of the salivatory centres associated with the facial and glossopharyngeal nuclei in the brainstem and (b) preganglionic sympathetic cells in the lateral horn of the upper thoracic spinal cord which synapse on postganglionic cells of the cervical sympathetic chain.

● **Efferent**. Stimuli pass to the parotid, submandibular, and sublingual salivary glands via (a) parasympathetic fibres of the **glossopharyngeal** (IX) and **facial** (VII) nerves which relay in cells of the otic and submandibular ganglia, respectively and (b) postganglionic fibres arising from cells of the cervical sympathetic chain. Activation of sympathetic and parasympathetic fibres increases production of saliva.

Swallowing reflex (7.8)

● **Afferent**. Swallowing is initiated voluntarily when the bolus of food or liquid is moved into the oropharynx and stimulates sensory branches of the **glossopharyngeal** (IX) **nerve**.

The information passes to the nucleus and tractus solitarius and spreads via interneurons to many cranial nerve motor nuclei including the nucleus ambiguus in the medulla which supplies most muscles of the palate, pharynx, and larynx.

● **Efferent** stimuli pass via motor nerves to the tongue, floor of the mouth, strap muscles, pharynx, and larynx. These raise and tense the soft palate and thus close off the nasopharynx. They raise the larynx, narrow the entrance to the larynx, and close the glottis. Activity in the pharyngeal plexus triggers a peristaltic wave which spreads from the pharynx downward through the oesophagus to the stomach.

If the oropharynx is stimulated, but not in the context of swallowing, then a **'gag' reflex** occurs in which motor fibres of the pharyngeal plexus cause a reflex contraction of the muscles of the pharynx, soft palate, and fauces. If the stimulus is sufficiently great, or is perceived to be so, then a relatively violent retching may occur which is accompanied by vomiting.

Yawn reflex (7.9)

● **Afferent.** Emotional stimuli associated with tiredness or those set up by the sight of others yawning pass from higher centres to respiratory centres in the medulla.

The stimulus spreads to somatic motor nuclei of the trigeminal (V) and facial (VII) nerves.

● **Efferent.** Stimulation of somatic motor fibres of the mandibular (V) and facial (VII) nerves results in the mouth being widely opened and activity in the phrenic nerves (C3, 4, 5) to the diaphragm and intercostal nerves (T1–12) to the external intercostal muscles causes a deep inspiration which is followed by a deep and often audible expiration.

Carotid sinus reflex

● **Afferent.** Changes in the degree of stretch within the walls of the carotid sinus stimulate fibres of the **carotid sinus nerve** (glossopharyngeal nerve; IX) which pass to the visceral sensory nucleus (nucleus solitarius) in the medulla.

The information spreads to preganglionic sympathetic cells in the lateral horn of the thoracic spinal cord which synapse on cells of the sympathetic chain or on sympathetic ganglia remote from the chain.

● **Efferent.** Postganglionic sympathetic fibres pass with spinal nerves or along blood vessels to innervate blood vessels and regulate blood flow and thus alter the peripheral resistance; in addition, sympathetic and parasympathetic innervation to the heart regulates the cardiac output so that the blood pressure is returned to normal.

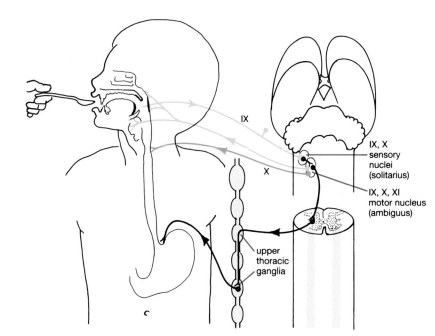

7.8
Swallowing and 'gag' reflex.

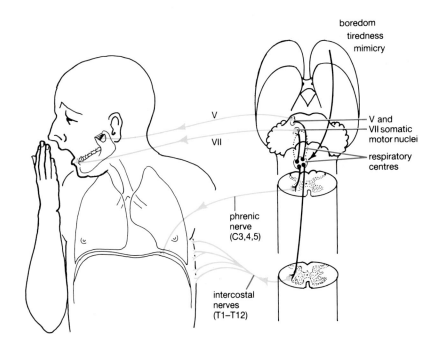

7.9
Yawning reflex.

CHAPTER 8

Answers to questions

Seminar 1

No questions.

Seminar 2

2A Extension of the head and neck is limited by tension in the anterior longitudinal ligament and by approximation of the spinous processes of the vertebrae.

2B Radiograph **6.2.7** shows a dislocation between the 5th and 6th cervical vertebrae.

2C Damage to sternomastoid or its nerve supply on one side will result in the head and neck being twisted toward the side of damage.

2D In the upper cavity of the joint, anteroposterior gliding movements occur which cause the head of the mandible to move forward when the mouth is opened widely. In the lower compartment, hinge-type movements occur between the mandible and the disc.

2E To reduce the dislocation it is necessary to press downward and posteriorly on the molar teeth of both sides to enable the head of the mandible and its disc to move backward over the articular eminence and re-enter the glenoid fossa.

2F The auriculotemporal nerve emerges behind the neck of the mandible and the chorda tympani branch of the facial nerve emerges from the squamotympanic fissure immediately behind the temporomandibular joint; either may be damaged.

2G These horizontally-orientated fibres are the only ones which retract the mandible.

2H Medial pterygoid helps to close the mouth; with lateral pterygoid it also protracts and causes side-to-side movement.

2I The medial and lateral pterygoids together protract the mandible.

Seminar 3

3A Cleft palate is often associated with an abnormally small mandible because, if the mandibular arch is small, the tongue may not descend at the appropriate time, and the palatal shelves fail to fuse.

3B Skin pigmentation masks skin colour due to the vasculature, but this can be judged instead by inspection of the conjunctiva and sclera.

3C When the eyes are gently closed, the two eyelids merely come together; when the eyes are 'screwed up', contraction of orbital fibres of orbicularis oculi pulls the lateral angle of the eye medially.

3D Buttress areas occur between: the pterygoid plates and the posterior aspect of the body of the maxilla; the zygoma and the lateral aspect of the supraorbital ridge of the frontal bone; and between the nasal process of the maxilla and the 'bridge' of the nose.

3E Damage to the facial nerve paralyses the muscles of facial expression. The eye cannot be closed properly and the conjunctiva may become dry and ulcerated; the mouth cannot be closed and 'dribbling' occurs at the angle; weakness in buccinator leads to the accumulation of food between the cheek and gums while eating.

3F Damage to the ophthalmic division of the trigeminal nerve causes loss of sensation over the conjunctiva and cornea, loss of the corneal reflex, and potential damage to the cornea.

3G If the cribriform plate is fractured, then CSF in addition to blood can leak from the nose.

3H Deviation of the septum can prevent proper drainage of the ethmoidal and maxillary sinuses.

3I 2nd premolar and molar roots often extend into the maxillary sinus so that root infection can lead to infection of the sinus.

3J A maxillary sinus can be drained and irrigated by directing a needle along the middle meatus and into the maxillary sinus through the lateral wall of the nose.

3K Bleeding often occurs from the cartilaginous part of the septum and can be stopped by sustained pressure of the soft lateral walls of the nostrils against the septum.

3L The walls of the nose are covered by a very vascular mucosa which warms the incoming air, hence the rich venous drainage.

3M The radiograph is angled to visualize the paranasal sinuses. The maxillary sinuses on both sides are opaque inferiorly and show fluid levels.

Seminar 4

4A The surface of the tongue is rough because it is studded with taste buds of different types.

4B The tongue can be moved in virtually every direction and this versatility is important during chewing, swallowing, and speech.

4C The acid in lemon juice stimulates taste buds, setting up a reflex which causes salivation.

4D In the floor of the mouth close to the opening of the submandibular and sublingual gland ducts.

4E The firm attachment protects it from shearing from the bone as a result of the forces generated by sucking and swallowing.

4F Touching the soft palate elicits a 'gag' reflex in which the palate is tensed and elevated, and the tongue and larynx pulled upward as in swallowing.

4G Hyoglossus pulls down the sides of the tongue; styloglossus pulls the tongue upward and backward.

4H The tongue can move backward and block the airway. The patient should be laid on his side and the tongue gently pulled forward by a finger in the mouth to keep the airway clear.

4I Because anastomoses between the inferior alveolar and lingual arteries are plentiful.

4J The swelling stretches the relatively non-expansile capsule, causing pain.

4K Every precaution, including the electrical stimulation of each piece of tissue about to be removed, is used to prevent parts of the facial nerve being damaged and the face paralysed.

4L Incisors are specialized for cutting, canines for gripping, and premolars and molars for grinding the food.

Seminar 5

5A The nasopharynx is closed off from the oropharynx.

5B They are elevated during swallowing and the laryngeal inlet closes.

5C As swallowing commences, the tip of the tongue is pressed against the hard palate. This helps to ensure that the bolus is ejected posteriorly into the oropharynx.

5D With the mouth open it is difficult for the tongue to be pressed against the hard palate, an important first component of swallowing.

5E The cannula must be guided beneath the inferior concha to the nasopharynx because the tubal opening is at the level of the inferior meatus.

5F The base of the sphenoid and occipital bones; the vomer forms the posterior part of the nasal septum.

5G The lymphoid tissue around the nasopharynx helps protect against bacterial invasion by this route.

5H A sharp body which penetrates the posterior wall of the laryngopharynx allows bacteria into the retropharyngeal space. The infection may spread downward into the posterior mediastinum and chest.

5I Buccinator arises from the anterior border of the pterygomandibular raphe (and from the maxilla and mandible opposite the molar teeth).

5J Salpingopharyngeus opens the pharyngotympanic tube and allows pressure to equilibrate between the nasopharynx and middle ear. This prevents discomfort, e.g. when ascending and descending in an aircraft.

5K The thoracic duct, which drains lymph from the thorax, abdomen, and lower limbs and from the left upper limb and left side of the head and neck, drains into the venous system at the junction of the left internal jugular and left subclavian veins.

5L (a) Styloglossus (XII) pulls the tongue upward and backward; (b) genioglossus (XII) thrusts the tongue against the hard palate while the posterior part of the tongue is pulled down by hyoglossus (XII); (c) tensor palati (mandibular V) and levator palati (pharyngeal plexus) tense and elevate the soft palate against the superior constrictor (pharyngeal plexus); (d) the inlet to the larynx is narrowed by the aryepiglottic muscle (recurrent laryngeal nerve, X).

Seminar 6

6A The laryngeal prominence is larger in men than women. This is a testosterone-dependent secondary sexual characteristic which appears at the time of puberty.

6B The stratified squamous epithelium which covers the vocal folds has a mechanical strength which protects against friction by the column of air. The epiglottis is also subject to friction on its upper surface and margins and is likewise covered with stratified squamous epithelium.

6C During swallowing the larynx is elevated by the suprahyoid and pharyngeal muscles.

6D The posterior wall of the trachea consists of muscle rather than cartilage. This arrangement provides for variation of the diameter of the trachea during quiet and forced respiration and for expansion when boluses of food pass down the oesophagus.

6E The upper margin of the quadrate membrane forms the aryepiglottic fold; the lower margin forms the vestibular fold.

6F The mucous membrane covering the vocal folds is closely attached to its free edge. Thus, swelling (due to inflammation or allergy) cannot disperse but will collect locally and can occlude the airway.

6G Mylohyoid, stylohyoid, and middle constrictor elevate the hyoid.

6H The recurrent laryngeal nerve supplies posterior cricoarytenoid and all other intrinsic muscles of the larynx except cricothyroid.

6I Cricothyroid tilts and pulls the thyroid cartilage forward on the cricoid cartilage, thus tensing the vocal fold.

6J In whispering, the intermembranous part of the glottis is adducted by lateral cricoarytenoid, and the intercartilaginous part abducted by contraction of posterior cricoarytenoid and relaxation of interarytenoid muscles.

6K Ectopic thyroid tissue can obstruct the airway if it grows large: (*a*) in the mouth; (*b*) where the thyroglossal duct loops up behind the hyoid bone; and (*c*) behind the manubrium, where enlargement can severely constrict the trachea.

6L If the lump moves upward on swallowing, then it is situated within, or firmly attached to the thyroid, pretracheal fascia, or larynx.

6M The intrinsic muscles of the larynx on the side of the damage will be paralysed, the voice will become hoarse, and the affected vocal fold will lie in the midline.

6N The arrows are pointing towards the internal jugular vein in the lateral part of the carotid sheath.

Seminar 7

7A Blood from a leaking aneurysm of an intracranial cerebral artery enters the subarachnoid space and could be detected in a sample of CSF taken by lumbar puncture. The bleeding might also raise CSF (intracranial) pressure; any gross increase in pressure must be excluded by examination of the retina before a lumbar puncture is performed.

7B Passive diffusion of CSF into cranial venous blood depends on: (*a*) higher hydrostatic pressure of the CSF; and (*b*) higher osmotic pressure of the venous blood due to the presence of plasma proteins. Pinocytosis also occurs.

7C A depressed anterior fontanelle is often associated with dehydration. A raised fontanelle indicates raised intracranial (CSF) pressure.

7D If an arteriovenous shunt occurred in the cavernous sinus, the immediate and most obvious effect would be swelling (oedema) of the orbital tissues which are normally drained by the sinus. The eye may protrude (exophthalmos) and even show signs of arterial pulsation.

7E The posterior pituitary is developed from the hypothalamus and is connected to the hypothalamus by the axons of neuroendocrine neurons; the anterior pituitary, by contrast, has a vascular link through the hypothalamo-hypophysial portal vessels.

7F Pituitary tumours commonly grow forward and upward to press on the optic chiasm. Optic nerve fibres arising from the nasal aspects of the retina, which cross the midline in the chiasm, are preferentially damaged, and loss of the temporal fields of vision (bitemporal hemianopsia; tunnel vision) results. Tumours may grow upward to press on the base of the hypothalamus, or downward to invade the sphenoidal air sinus.

7G Traction on the abducent nerve(s) would lead to paralysis of the lateral rectus muscle(s) which rotates the gaze laterally, and therefore would cause a medial squint.

7H The trigeminal ganglion contains the cell bodies of trigeminal sensory neurons (except those serving proprioception, which are located in the mesencephalic nucleus of the trigeminal nerve).

7I Blood leaking from a ruptured middle meningeal artery would collect in the extradural space (localized by the close attachment of the dura to the skull sutures) and raise intracranial pressure. An extradural haematoma in the region of the pterion can press on the motor speech area (if on the 'dominant' side) and cause slurred speech; it may also irritate the motor area for the fingers and cause small involuntary movements of the fingers on the opposite side. A haematoma derived from the posterior branch can press on the auditory cortex.

7J Pus from a frontal sinus which erupted through the frontal bone and dura would also penetrate the delicate arachnoid mater and enter the subarachnoid space, causing inflammation of the meninges (meningitis). It would track along either side of the falx cerebri, and this can be visualized on coronal CT.

7K The nerves most likely to be injured in the fracture of the base of the skull shown in **6.7.10** are the right facial (VII) and vestibulocochlear (VIII) nerves.

Seminar 8

8A White sclera can be seen both below and above the cornea and iris. This is abnormal (exophthalmos) and likely to be due to an increase in the orbital tissue behind the eyeball, causing it to protrude.

8B Absence of the blinking reflex can result in damage and ulceration of the cornea.

8C The pupil of the opposite side should also constrict in response to a light stimulus, because the neural pathways serving the light reflex are bilaterally connected.

8D If the eyeball is too long, objects at infinity cannot be focused by the lens on to the retina and short sight (myopia) results; if the eyeball is too short, even maximal lens curvature will not focus nearby objects and 'long sight' (hypermetropia) develops.

8E Continued compression on the veins will lead to increasing pressure within the optic nerve and eventual blindness. The optic disc will appear blurred at the edges and white at its centre owing to atrophy of optic nerve fibres.

8F When contact lenses are worn most refraction occurs at the air/lens interface.

8G Failure of the aqueous humour to drain leads to an increased volume and therefore pressure in the anterior chamber of the eye, which is transmitted throughout the eyeball. This causes pain and subsequent damage to the retina, with blindness resulting if the condition (glaucoma) is not diagnosed and treated rapidly. The irido-corneal angle can be widened by constriction of the pupil, so that cholomimetic drugs may be helpful.

8H When the lens becomes stiff, the power of accommodation is first diminished and then lost, so that spectacles of different focal lengths are needed to focus near and distant objects.

8I Disruption of the attachments of the suspensory ligament of the eyeball and depression of the eyeball in the orbit leads to 'double vision' (diplopia) because the brain is unable to fuse the two overlapping images which are no longer in register.

8J The pull of superior and inferior rectus passes medial to the vertical axis of the eye so that these muscles, respectively, cause upward and medial, or downward and medial movement of the cornea.

8K Contraction of inferior oblique rotates the cornea upward and laterally.

8L If the oculomotor nerve is damaged, the eyelid cannot be elevated and the eye remains closed. If the sympathetic fibres only are damaged (Horner's syndrome), then the eyelid droops abnormally, but the eye is not completely closed.

8M Damage to the oculomotor nerve on one side leads to ptosis (levator palpebrae), lateral deviation of the eye (unopposed action of lateral rectus and superior oblique), inability to move the eye medially (medial, superior, and inferior recti) or upward and outward (inferior oblique). The pupil would not constrict in response to light or accommodation (parasympathetic fibres to sphincter pupillae).

8N Damage to the facial nerve which paralyses orbicularis oculi prevents the eyes from being properly closed; the conjunctiva and cornea become dry and therefore liable to infection, ulceration, and opacity.

8O The nasolacrimal duct drains into the inferior meatus of the nose.

Seminar 9

9A Cupping a hand behind the ear increases the effective size and thus sound-gathering ability of the pinna. This can be aided by turning the head so that one ear is oriented directly towards the sound source.

9B Cartilage has no *intrinsic* blood supply so that, when trauma to the ear causes bruising, the skin and its blood vessels are lifted away from the cartilage, which suffers avascular damage.

9C Where infection occurs, inflammation causes an increase in tissue fluid which usually spreads into the surrounding tissues. Where the skin is firmly attached to periosteum or perichondrium the fluid cannot disperse and the tissues are tightly stretched, causing pain.

9D When flying at a high altitude the cabin pressure is usually below atmospheric pressure on the ground and the pressure within the middle ear becomes equilibrated via the pharyngotympanic tube, which is open only on occasion. During descent, cabin pressure increases and the tympanic membrane is forced inward, creating tension and the sensation of 'bursting'. Swallowing causes pharyngeal muscles attached to the cartilaginous part of the pharyngotympanic tube (salpingopharyngeus) to open the tube and allow equilibration of pressure.

9E Tensor tympani and stapedius are both involved in reflexes which attenuate, respectively, low- and all-frequency oscillations of the ossicle chain. During speech they attenuate sound input to the cochlea from the speaker's own larynx and are activated before vocalization starts. They contract after a loud external sound arrives, but protect the inner ear during prolonged loud noises.

9F Fractures of the base of the middle fossa of the skull often involve the roof of the middle ear cavity and venous sinuses running along the petrous temporal bone; they may also rupture the internal jugular vein or internal carotid artery, which bleed through the floor. Blood in the middle ear appears in the external auditory meatus through the (ruptured) tympanic membrane.

9G Infection in the middle ear can spread into the mastoid air cells, and can penetrate through the thin bony roof to the CSF in the subarachnoid space (meningitis), the venous sinuses lying on the petrous temporal bone, and even the temporal lobe or cerebellum.

9H Endolymph contains a relatively high potassium concentration (144 mм l^{-1}) and low sodium concentration (16 mм l^{-1}) compared to perilymph (potassium 5 mм l^{-1}; sodium 150 mм l^{-1}).

9I Stimuli to the external auditory meatus activate vagal reflexes, which include increased peristalsis. But beware, children may vomit if the external auditory meatus is examined roughly, and aged persons with heart conditions have suffered fatal vagal slowing of the heart as a result of minor procedures such as syringeing wax from the ears.

Seminar 10

10A As the head is rotated, sternomastoid comes to lie over the carotid sheath. In attempting to sever the carotid/internal jugular vessels an intending suicide often extends and rotates the head to expose more of the neck, thus bringing sternomastoid over the vessels and protecting them.

10B The inferior thyroid artery divides into a series of branches before it enters the lower pole of the gland; the branches intermingle with the recurrent laryngeal nerve as it ascends to the larynx. The nerve must therefore be identified and preserved during thyroidectomy operations.

10C Temporary occlusion of the vertebrobasilar arteries causes loss of brainstem, cerebellar, and occipital lobe function which leads, *inter alia*, to disturbances of equilibrium and vision.

10D Occlusion of the ophthalmic artery causes blindness because the central artery to the retina is an end-artery. Other components of the orbit would probably survive because of the anastomoses made by their supplying vessels.

10E The cerebral cortex of the central nervous system uses oxygen more rapidly than other tissue and, therefore, any deficit in supply affects it first. The internal carotid artery supplies parts of the cerebral cortex which are involved with movement and sensation other than vision. A deficit in the internal carotid supply can therefore cause paralysis and/or disorders of sensation.

10F When the neck is rotated, the face and mouth move considerably with respect to the neck and the carotid tree; likewise, in swallowing, the pharynx and larynx must be able to move up and down; the arterial loops permit this movement.

10G Because of the extensive anastomosis between the labial branches of opposite sides, bleeding can be arrested by direct pressure, or compression of both the left and right facial arteries as they cross the mandible.

10H The anastomoses are shown on **6.10.7**.

10I The pressure of blood in the intracranial venous sinuses will depend on the position of the body; in the upright position the hydrostatic pressure would be negative.

10J The hydrostatic pressure of the CSF is higher than that of venous blood in the sinus because it is actively secreted. The osmotic pressure is lower because the CSF contains less protein and glucose than venous blood.

10K The carotid sheath is thinner over the vein because it must be able to expand to accommodate increased flow.

10L Enlarged adenoids can block the internal nares of the nose, so that the child persistently breathes through the mouth.

10M (*a*) The posterior part of the scalp; (*b*) the anterior aspect of the mouth and tongue.

10N The internal carotid artery is occluded at its source. Paralysis and sensory loss of the opposite side of the body are likely to result from the lack of blood supply to this side of the cerebrum.

10O In this combined arteriogram of the vertebral and internal carotid arteries, the middle cerebral artery and its branches have not been filled with contrast material; thus there is a block at the origin of the middle cerebral artery.

Seminar 11

11A The nose obtrudes into the visual field on the medial side.

11B The abducent nerve (VI) supplies lateral rectus which moves the gaze laterally; the trochlear nerve (IV) supplies superior oblique which moves the gaze downward and laterally; the oculomotor nerve supplies the remaining extraocular muscles, including levator palpebrae, and all those that move the gaze medially.

11C The anterior cranial fossa.

11D A defect in both temporal fields of vision is usually caused by damage to the nasal retinal fibres as they cross in the optic chiasm, usually as a result of an upward growing tumour of the pituitary gland.

11E (*a*) Blindness in the right eye; (*b*) a defect in the left visual field (detectable by testing either eye) (see **6.11.2**).

11F The ptosis is due to paralysis of levator palpebrae; the divergence to paralysis of medial, superior, and inferior rectus; the pupillary dilatation to loss of parasympathetic supply to sphincter pupillae.

11G Damage to the trochlear nerve paralyses superior oblique. When looking on or above the horizontal plane the affected person has normal vision, but on looking downward the eyeball becomes rotated and double vision (diplopia) results.

11H Damage to the abducent nerve causes paralysis of lateral rectus and therefore a medial (convergent) squint, with diplopia.

11I The vesicles would be distributed over the forehead to the vertex, over the upper eyelids, conjunctiva (particularly serious), and in the nose.

Seminar 12

12A If the sound is heard when conducted through bone but not air, then the external ear, tympanic membrane, ossicle chain, or fenestra ovalis are failing to transmit the sound vibrations to the cochlea.

12B If the maxillary sinus is inflamed, the superior alveolar nerves may become affected and pain may be experienced in the upper teeth.

12C Damage to the parasympathetic nerve supply of the lacrimal gland reduces lacrimal secretion, which leads to inadequate moistening of cornea and conjunctiva, a reddened eye, and ulceration. A facial injury which fractures the zygomatic bone can damage the parasympathetic fibres in the zygomatic nerve *en route* to the lacrimal gland, or the preganglionic supply may be cut off by a lesion of the facial nerve.

12D The trigeminal fibres of the lingual nerve supply touch and pain sensation to the anterior two-thirds of the tongue and the floor of the mouth; sensory fibres of the chorda tympani (which joins the lingual nerve) supply taste sensation.

12E The lingual nerve runs close to the inferior alveolar nerve at this point; therefore both touch and taste sensation in the anterior 2/3 of the tongue are commonly lost in addition to the desired anaesthesia of the lower teeth and gums; and an area of skin on the chin may also be anaesthetized (mental nerve).

12F The postganglionic parasympathetic secretomotor fibres from the pterygopalatine ganglion supply the lacrimal gland, and mucous glands in the nose, nasopharynx, and palate.

12G Facial muscles are formed from mesoderm of the 2nd branchial arch and migrate to their destinations.

12H Pressure on the facial nerve in the stylomastoid canal paralyses the muscles of facial expression including buccinator and platysma. Tears and saliva would overflow the corner of the eye and mouth, respectively, and food would tend to lodge in the vestibule of the mouth between the cheek and the teeth.

12I In addition to paralysis of facial muscles, damage to the facial nerve in the internal auditory meatus causes loss of lacrimation, loss of taste from the anterior two-thirds of the tongue, and hyperacusis (increased perception of sound) on the affected side.

12J Cochlear nerve cell bodies form the spiral ganglion in the modiolus of the cochlea; vestibular cell bodies form the vestibular ganglion in the depths of the internal acoustic meatus.

Seminar 13

13A The spinal accessory nerve can be tested through the function of sternomastoid — forced rotation of the head — or trapezius — shrugging the shoulders.

13B The postganglionic secretomotor fibres join the auriculotemporal nerve for distribution to the parotid gland.

13C Stimulation of the external auditory meatus affects the vagus nerve which causes slowing of the heart.

13D The internal laryngeal nerve pierces the thyrohyoid membrane to enter the larynx.

13E The right recurrent laryngeal nerve loops beneath the right subclavian artery.

13F The voice would be hoarse; inspection of the larynx would reveal a paralysed left vocal fold lying with its free edge at about the midline.

13G (a) A failure of one-half of the soft palate to elevate when saying 'Aah'; (b) weakness of trapezius on the affected side; (c) weakness of sternomastoid on the affected side.

13H The right side of the tongue is paralysed and will eventually atrophy; on attempting to protrude the tongue, it deviates to the affected side.

Seminar 14

14A All spinal function below C3 would be lost (i.e. paralysis and anaesthesia of limbs, thorax, and abdomen and pelvis). Since the diaphragm intercostal muscles would be paralysed, the patient would require permanent artificial respiration.

14B Postganglionic sympathetic fibres reach the submandibular and otic ganglia from the external carotid plexus along the lingual artery and maxillary artery, respectively; they are distributed with branches of the ganglia.

14C The partial ptosis results from loss of sympathetic innervation to part of levator palpebrae; the dry, flushed skin from the absence of sympathetic control of sweating and vasoconstriction; the pupillary constriction from the paralysis of dilator pupillae which leaves sphincter pupillae unopposed; the enophthalmos is largely apparent and due to the ptosis.

14D For the branches of the subclavian artery see p. 00.

14E T1 is likely to be affected, with weakness and wasting of the small muscles of the hand and anaesthesia over the inner aspect of the arm; if the phrenic nerve is involved, one dome of the diaphragm will be paralysed; if the recurrent laryngeal branches of the vagus are involved, the voice will become hoarse.

14F A cervical rib is visible on the left side of the radiograph. The T1 root of the brachial plexus crosses over a cervical rib (or the fibrous band that joins it to the 1st rib), and the nerve can become injured, especially when the arm is pulled downward as by carrying a heavy weight. This can lead to weakness of the small muscles of the hand which T1 supplies.

Index